Climate Change in Prehistory

The End of the Reign of Chaos

Climate Change in Prehistory explores the challenges that faced humankind in a glacial climate and the opportunities that arose when the climate improved dramatically around 10,000 years ago. Drawing on recent advances in genetic mapping, it presents the latest thinking on how the fluctuations during the ice age defined the development and spread of modern humans across the Earth. It reviews the aspects of our physiology, intellectual development and social behaviour that have been influenced by climatic factors, and how features of our lives – diet, health and the relationship with nature – are also the product of the climate in which we evolved. This analysis is based on the proposition that essential features of modern societies – agriculture and urban life – *only* became possible when the climate settled down after the chaos of the last ice age. In short: climate change in prehistory has in so many ways made us what we are today.

Climate Change in Prehistory weaves together studies of the climate with anthropological, archaeological and historical studies, and will fascinate all those interested in the effects of climate on human development and history.

After seven years at the UK National Physical Laboratory researching atmospheric physics, Bill Burroughs spent three years as a UK Scientific Attaché in Washington DC. Between 1974 and 1995, he held a series of senior posts in the UK Departments of Energy and then Health. He is now a professional science writer and has published several books on various aspects of weather and climate (two as a co-author), and also three books for children on lasers. These books include *Watching the World's Weather* (1991), *Weather Cycles: Real or Imaginary* (1992; second edition 2003), *Does the Weather Really Matter?* (1997), *The Climate Revealed* (1999), and *Climate Change: A Multidisciplinary Approach* (2001), all with Cambridge University Press. In addition, he acted as lead author for the World Meteorological Organization on a book entitled *Climate: Into the Twenty-First Century* (2003, Cambridge University Press). He has also written widely on the weather and climate in newspapers and popular magazines.

Climate Change in Prehistory

The End of the Reign of Chaos

WILLIAM JAMES BURROUGHS

CAMBRIDGE UNIVERSITY PRESS
Cambridge, New York, Melbourne, Madrid, Cape Town, Singapore, São Paulo

CAMBRIDGE UNIVERSITY PRESS
The Edinburgh Building, Cambridge CB2 2RU, UK

Published in the United States of America by Cambridge University Press,
New York

www.cambridge.org
Information on this title: www.cambridge.org/9780521824095

First published 2005

Printed in the United Kingdom at the University Press, Cambridge

A catalogue record for this book is available from the British Library

ISBN-13 978-0-521-82409-5 hardback
ISBN-10 0-521-82409-5 hardback

Contents

Preface

Gazing up at the up at the roof of the reconstruction of the cave at Lascaux in southwestern France, it is a stunning realisation that the magnificent paintings were drawn some 17 000 years ago. Sometimes referred to as the 'Sistine Chapel of Prehistory', this artistic marvel was painted at a time when the northern hemisphere was about to emerge from the steely grip of the ice age. This sense of wonderment is compounded by the knowledge that the more recent discovery of similar paintings in the Chauvet cave, in the Ardeche region of France, has been dated as much as 15 000 years earlier. So, more than 10 000 years before the first recognised civilisations of Mesopotamia and Egypt emerged, over many thousands of years, the ice-age hunters of Europe were producing these extraordinary examples of creativity.

Confronted by so much talent so long ago a stream of questions arises. Where did these people come from? Where did they go? What were conditions like at the time? What happened to the skills they had developed? Did the changes in the climate that followed explain why they faded from view? What happened to the skills they had developed? What were the consequences of this apparently frustrated development?

Answers to these questions, and many more, are starting to emerge from two areas of science that have transformed our understanding of the development of humankind in prehistory. First, we can draw on advances in climate change studies of recent decades. Measurements of samples from tropical corals to Greenland's icy wastes, from sediments at the bottom of the world's oceans and lakes, from stalactites and stalagmites deep in the bowels of the Earth, and from living and long-dead trees have transformed our understanding of how the climate has changed in the past. These advances provide a detailed picture of the chaotic climate of the ice-age world, which threatened the very existence of our species. How our ancestors survived these challenges is a vital part of our history.

The other scientific development is, in some ways, even more extraordinary. By unravelling the information that is locked up in our DNA, we can address the deeply personal question of how are we linked to the people who survived the ice age. This contains a record of the entire evolution of humankind. Although there are limitations to what we can find out, two things are central to unlocking the secrets in our genes. The first is a statement of the obvious. This is that not a single one of our own direct ancestors died without issue. So there is an unbroken genetic line from all of us to people living during the ice age. In addition, while there is no way of knowing precisely where our own ancestors were living then, the whole new world of genetic mapping can tell us an amazing amount about our origins. This includes a variety of insights into how modern humans peopled the world and how this process was influenced by climate change.

My aim in this book is to describe how scientific advances have opened up new perspectives of the evolution of humankind in a world where climatic chaos was the norm. It will take us into many aspects of the lives of our ancestors and those of the creatures living around them, and explore how overcoming the challenges of the ice age made us what we are today.

Acknowledgements

Because this book draws on lengthy personal involvement in climate matters, it is difficult to identify all the people who have helped me to form a view on the many facets of climate, how it has changed and its impact on all our lives. Among the meteorological community I would like to thank Chris Folland, David Parker, John Mitchell and Bruce Callendar at the UK Meteorological Office, David Anderson, Tim Palmer, Tony Hollingsworth, and Austen Woods at the European Center for Medium Range Weather Forecasting, Grant Bigg, Keith Briffa, Mike Hulme and Phil Jones at the University of East Anglia, John Harries and Joanna Haigh at Imperial College, and Tom Karl at NOAA Climate Data Center, for helpful discussions, and the provision of data and other material, which in one way or another were essential for completion of the book. I am grateful also to Richard Alley, Michael Mann, Martin Parry, Julia Slingo, Tony Slingo and Alan Thorpe for helpful advice on climate matters. In addition I am most grateful to Tony Barnosky, Clive Bonsall, Jean Clottes, Francois Djindji, Ed Hollux, Sharon Kefford, Nigel Speight, Michael Vellinga and Uli von Grafenstein who provided advice, pictorial material, or both.

I would also like to acknowledge the modern practice of making data accessible on the Internet. Whether it is individual workers, or the large teams that lie behind many of the major research efforts producing the results that are reviewed here, without the spirit of openness and sharing it would have been much more difficult to produce this book. This is particularly true of new developments involving large, often multinational teams. So, I am particularly grateful for the accessibility of data on websites such as the World Data Center for Paleoclimatology, Boulder, Colorado, USA, and also for individual sites where researchers have made their results available. I hope, where I have used material, I have adequately acknowledged the original source.

Finally, I am deeply indebted to my wife who, as always, helped and supported me throughout the lengthy gestation of this book.

1 Introduction

Chaos umpire sits,
And by decision more embroils the fray
By which he reigns: next him high arbiter
Chance governs all.

John Milton (1608–1674), *Paradise Lost*

There is a cosy notion that progress is a natural consequence of the development of human social structures. Reinforced by the rise of Europe from the Middle Ages and the subsequent exploitation of the New World, it is all too easy to forget past setbacks. 'Dark Ages' have punctuated the recorded history of our species. The period following the decline and fall of the Roman Empire is probably the best-known example, but sudden and catastrophic declines of earlier ancient civilisations are important reminders that progress is not an automatic part of the human condition. In popular culture this simple onward and upward view of human development extends back into the Palaeolithic: as the Earth gradually emerged from the ice age the human race stumbled from its caves and started its ascent to civilisation as we know it. While this is a parody of our current understanding about what really happened, it still lurks deep within our cultural subconscious. What it loses sight of is the extent of intellectual development that had been established in prehistory (Rudgley, 1998). In some instances discoveries were made independently at different places and at different times. These punctuated developments may have been a consequence of climatic events, and this tortuous process is part of the story explored.

Surviving the rigours of the ice age also profoundly influenced the evolution of modern humans. The fluctuations within this glacial period dictated how we spread out across the globe. They are hard-wired into us in respect of our genes, stature and health, and integrated into our attitudes to gender, warfare, animals and much more.

In exploring so many aspects of human life there is further complication. A surprising number of the areas of scientific research discussed here involve bitter academic feuds. At every turn throughout this extended interdisciplinary discussion we will find highly respected professionals slugging it out in august journals. The objective of the book is to present a balanced account of how the various debates fit into the wider picture, always recognising that this is a matter of tiptoeing through a series of intellectual minefields.

1.1 CAVE PAINTINGS

In the context of understanding prehistory, and how climate change, in particular, played a part in stimulating progress or bringing it to a grinding halt, several developments in the 1990s acted as the inspiration for this book. The first was the discovery of the images found in the Chauvet cave in 1994. When carbon dating (see Appendix) of the charcoal used in these breath-takingly beautiful drawings of animals showed that they were over 30 000 years old, the archaeological world was taken aback (Clottes *et al.*, 1995).

This dating was some 15 000 years (15 kyr) earlier than had been expected, as the images bore a striking resemblance to the much better known drawings in the caves in Lascaux and Altimira that date back to around 17 000 years ago (17 kya). So rather than being the product of the developments that were seen as part of Europe emerging from the last ice age, these images were drawn by our forebears whose descendants had yet to survive the extreme stages of the last ice age, which plunged all of Europe north of the Alps and the Pyrenees into cold storage for over 10 kyr. The only significant difference in the images was that those from the earlier era depicted a world inhabited by more dangerous animals. In particular, the many images of lions (Fig. 1.1) are something that rarely appears in later artwork.

Inevitably, the question of the validity of the dating was raised. These doubts took time to address. In addition, the sensational nature of the Chauvet discovery diverted attention from the growing evidence of a much longer artistic tradition in Europe. In defending the

FIGURE 1.1 A painting of lions from the Chauvet cave, which has been dated as being over 30 000 years old. (With the kind permission of Jean Clottes.)

Chauvet dates, improved measurements were obtained from a number of other French caves (Valladas *et al.*, 2001), including recent exciting discoveries at Cosquer and Cussac. This analysis confirmed what was becoming increasingly evident from a wide range of sites that palaeolithic cave art was part of an artistic continuum dating back to before 30 kya.

The draughting skills of the people who had created these images transfix the viewer. Any frustrated artist, who has struggled to master the essentials on line and weight in drawing, can only genuflect to someone who had got it in one. This reaction was encapsulated in the earlier observation of an artist who had these skills in abundance – Pablo Picasso – who, on seeing the paintings in the Lascaux cave, observed, 'We have invented nothing!'

What are the implications of these skills surviving for so long? The oldest dates found so far are in the Chauvet cave (between 32 and 30 kya), while the most recent are found in the cave at Le Portel (11.6 kya; Clottes, 2002). Stop and think just how long this period is. It is some 800 generations, or more than ten times the period since the fall of the Roman Empire. This immense period of time suggests that in order for such a tradition to persist, there must have been an effective form of passing on this knowledge. Without it, the fundamental unity of this art could not have survived for so long. Possibly of even greater importance is that the assumption that was made before Chauvet was discovered – that the evolution of art had been gradual, from coarse beginnings rising to an apogee at Lascaux – cannot be sustained. The recent discoveries have shown that as early 30 kya sophisticated artistic skills had already been invented. So, even if the exercise of these skills lapsed from time to time, there was a social consciousness that enabled them to be sustained.

This sense of continuity raises fascinating questions about what was happening to the world during this immense period of time. Here we have the benefit of a second more consequential scientific development. Since the 1960s scientists have been drilling ice cores and making measurements of their properties. Snow deposited on the ice sheets of Antarctica and Greenland, and in glaciers in mountain ranges around the world, contains a remarkable range of information about the climate at the time it fell. Where there is no appreciable melting in summer, the accumulation of snow, which is compressed to form ice, contains a continuous record of various aspects of climate variability and climate change. This includes evidence of changes in temperature, the amount of snow that fell each year, the amount of dust transported from lower latitudes, fall-out from major volcanoes, the composition of air bubbles trapped in the ice and variations in solar activity. The best results are, however, restricted to Antarctica and Greenland, with more limited results from glaciers and ice caps elsewhere around the world.

The dramatic advance with ice cores came with the publication in the early 1990s of the first results of two major international projects: the Greenland Ice Sheet Project Two (GISP2) (Grootes *et al.*, 1993) which successfully completed drilling a 3053-m-long ice core down to the bedrock in the Summit region of central Greenland in July 1993; and its European companion project, the Greenland Ice Core Project (GRIP) (Greenland Ice Core Project Members, 1993), which one year earlier penetrated the ice sheet to a depth of 3029 m, 30 km to the east of GISP2. These cores provided a completely new picture of the chaotic climate throughout the last ice age, the turbulent changes that occurred at the end of this glacial period and the stability of the climate during the last 10 kyr (a period known as the Holocene).

These chaotic changes were evident in many of the ice-core parameters, including rapid fluctuations in the snowfall from year to year and sudden changes in the amount of dust swept up from lower latitudes. The most spectacular results were obtained, however, by measuring the ratio of oxygen isotopes (oxygen-16 and -18), which provided an accurate record of regional temperature over the entire length of the ice core. The amount of the heavy kind of oxygen atoms, oxygen-18 (^{18}O), compared with the lighter far more common isotope oxygen-16 (^{16}O), is a measure of the temperature involved in the precipitation processes. But this is not a simple process. The snow is formed from water vapour that evaporates from oceans at lower latitudes and travels to higher latitudes. The water molecules containing ^{16}O are lighter, and evaporate slightly more readily and are a little less likely to be precipitated in snowflakes than those containing ^{18}O. Both effects are related to the temperature, so the warmer the oceans and the warmer the air over the ice caps the higher the proportion of ^{16}O in the snow that fell. So during warm episodes in the global climate the proportion of the ^{18}O in the ice core is lower.

These cores presented an entirely different picture of the climate during and following the last ice age. Added to the glacial slowness of changes that led to the building and decline of the huge ice sheets was a whole new array of dramatic changes (Fig. 1.2). While these long-term consequences remained, two exciting features emerged from the detailed record of the ice cores. First, they provided much improved evidence of the frequent fluctuations in the climate on the timescales of millennia that ranged from periods of intense cold to times of relative warmth. Second, and even more interesting, these longer-term variations were overlain with evidence of dramatic short-term fluctuations: over Greenland, annual average temperatures rose and fell by up to 10 °C in just a few years, while annual snowfall trebled or

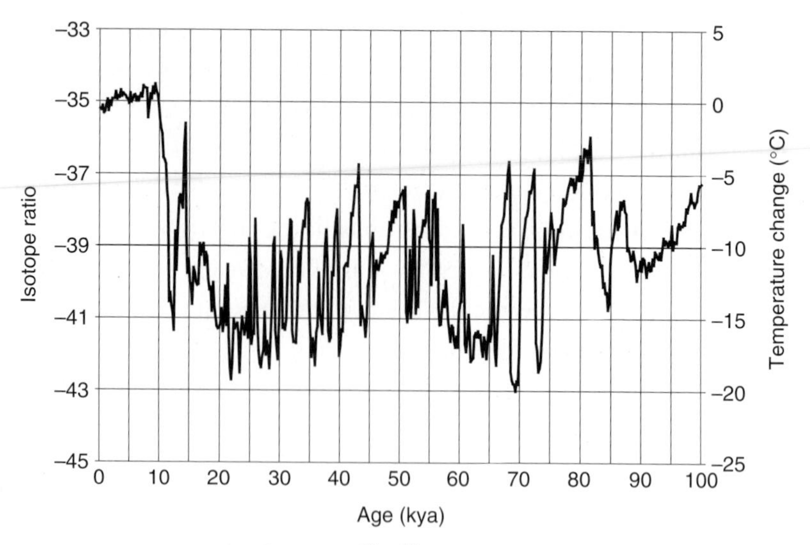

FIGURE 1.2 The changes in $^{18}O/^{16}O$ isotope ratio observed in the GRIP ice core for 200-year intervals over the last 100 000 years (0 to 100 kya), together with an approximate estimate of the changes in temperature that have taken place over this period. (Data archived at the World Data Center for Paleoclimatology, Boulder, Colorado, USA.)

declined by a third. As the research team memorably described the patterns (Taylor *et al.*, 1993), the climate across the North Atlantic behaved like a 'flickering switch'.

As for looking farther back in time, just how much could be extracted from ice cores became clear from work in Antarctica. Results obtained from high on the ice sheet, by the European Project for Ice Coring in Antarctica (EPICA) (Fig. 1.3), extended records back through more than 730 kyr, covering eight ice ages (EPICA Community Members, 2004). When combined with earlier ice-core data and other records obtained from around world from ocean sediments, lakebeds and peat deposits, and stalactites and stalagmites, these now give us a remarkably detailed picture of the climate of the last few hundred thousand years. This is offering archaeologists the opportunity to look with far greater precision at the conditions that controlled the development of humankind during the last ice age and the warming that followed it.

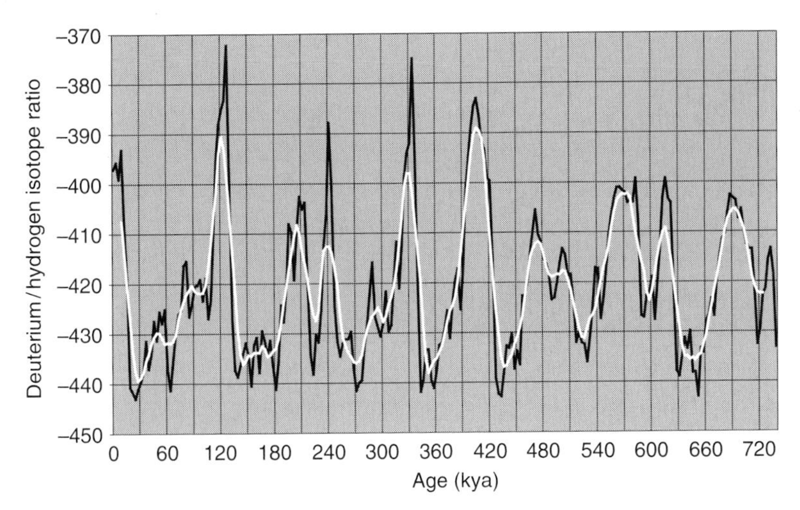

FIGURE 1.3 The measurement of the deuterium/hydrogen isotope ratio (D/H) in an ice core drilled by the European Project for Ice Coring in Antarctica (EPICA) at Dome C in Antarctica, showing how the ice-core record extends back 736 000 years (736 kyr) covering the last eight glacial cycles. The black line is the average values for every 3000 years and the white line is the seven-point running mean of these data. (Data from EPICA community members (2004), supplementary information, www.nature.com/nature.)

The impact of climate change on social and economic development has been a part of historical analysis for many years. There is a tendency to think that human capacity to create the intellectual accomplishments that are so much a part of recorded history did not really blossom until the establishment of identifiable civilisations. In fact, the kernel of these processes formed while grappling with the hardships of the ice age. The evidence of cave paintings and other forms of artistic activity suggests that these intellectual capacities were well developed long before agriculture and the establishment of sizeable human settlements (Rudgley, 1998). It shows clearly that the desire to record accurately observations about the world around us, and to pass that information on to both contemporaries and subsequent generations, dates back to these times. That these images also exhibit sublime aesthetic and spiritual components resonates even more with our own experience.

These images contain something of ourselves and the ideas we seek to represent in art. This does not mean that our distant ancestors produced these pictures for the same reasons that we might today or that we are capable of explaining what their purpose was. Indeed, the danger of inflicting our current perceptions on their imagery takes us into complicated social and psychological analysis (Lewis-Williams, 2002, pp. 41–68). In particular, we run the risk of seeking to impose on palaeolithic ancestors our contemporary concerns about sex, social equality and gender roles. As one writer memorably observed, 'palaeolithic art has often been a "Rorschach [inkblot] Test" in that modern-day observers have tried to read into the mind and spirit of primitive humans, but they have perhaps learned more about their own psyches than about the primitives' (Wenke, 1999, p. 209). For the moment, all that needs to be said is that they already had highly developed understandings of the flora and fauna around them, and were superb draughtsmen or draughtswomen.

The fundamental question is how these creative features of the minds of humans so long ago, with which we can so readily identify, helped them overcome the challenges of the worst of the ice age. These intellectual capabilities were an integral component of their survival. They influenced how they evolved through the long dark night of the ice age and the chaotic dawn of the Holocene. Furthermore, the fact that there is a thread that links us to these people provides particular insight into how modern humans were able to seize the opportunities presented by the climatic amelioration when the ice age ended.

1.2 DNA SEQUENCING

Another development of recent decades has transformed our thinking about human prehistory. This is the whole new science of genetic mapping that brings an entirely different perspective to our past. The discovery of the structure of DNA 50 years ago has altered how we view human evolution. It established the amazing concept that each of us, within our DNA, has a record of the entire evolution of humankind. The development of rapid and inexpensive methods of probing DNA sequences has led, since around 1980, to its application to evolutionary studies and to the creation of the subject of molecular anthropology. The most complete way of sequencing the human genome is to determine the exact order of the 3 billion chemical building blocks (called bases and abbreviated A, T, C and G) that make up the DNA of the 24 different human chromosomes. The sequencing of the entire genome, which contains some 30 000 genes, is the central challenge in the Human Genome Project.

The essential feature of DNA is that it carries the replication instructions for cell division. This process is carried out with extreme reliability, but in about one in a billion divisions mutations occur. This has the consequence that between successive generations these mutations slowly accumulate in the DNA of any species. It is the accumulation of mutations in the DNA that provides both the grist in the mill of natural selection and the metronome underlying the molecular clock. By comparing DNA sequences and measuring the incidence of genetic markers in human and animal populations around the world, it is possible to draw conclusions about the timing of separation of different species and different groups of humans.[1]

If this process required the sequencing of the entire genome and examining changes in all 30 000 genes it would be impossibly difficult. The breakthrough in the mapping process has come with the discovery that by sequencing two specific parts of human genome it is possible to explore the slow ticking of the human molecular clock. The first of these is mitochondrial DNA (mtDNA; for this and other technical terms, see the Glossary), which is a section that is some 16 000 base pairs long. This is only passed through the female line of the species and also has the benefit that mutations occur more rapidly there than elsewhere in the genome. Nevertheless, the process is very slow. If two people had a common maternal ancestor 10 000 years ago,

[1] An introduction to the subject of genetic mapping can be found in such books as Cavalli-Sforza (2000), Sykes (2002) and Oppenheimer (2003) (see Bibliography).

then there would be one difference in their genetic sequences. The parallel development has been to study the differences in the Y-chromosome, the only purpose of which is to create males.

Studying the genetic variation of mtDNA and the Y-chromosome across human populations can produce objective data that provide new insights into human history over many thousands of years, such as the colonisation of previously uninhabited areas and subsequent migrations. Prior to this the only equivalent analysis relied on the study of languages to infer a pattern of human development. This work, sometimes termed *glottochronology*, cannot delve as far back into the past as genetic studies, and will not be considered in any detail in this book. Here we will concentrate on genetic mapping, which provides an independent picture of how modern humans peopled the world during and after the last ice age. This analysis must, however, be combined with other sources of knowledge, such as archaeological or historical records, to form a balanced view.

1.3 ARCHAEOLOGICAL FOUNDATIONS

The next stage in this introduction is to confront the challenge of archaeology. It is often easier to write with confidence on fast-developing and relatively new areas of research, such as climate change and genetic mapping, than to review the implications of such new developments for a mature discipline like archaeology. Because the latter consists of an immensely complicated edifice that has been built up over a long time by the painstaking accumulation of fragmentary evidence from a vast array of sources, it is hard to define those aspects of the subject that are most affected by results obtained in a completely different discipline. Furthermore, when it comes to many aspects of prehistory, the field is full of controversy, into which the new data are not easily introduced. As a consequence, there is an inevitable tendency to gloss over these pitfalls and rely on secondary and even tertiary literature to provide an accessible backdrop against which new developments can be more easily projected.

In adopting this cautious approach there is a risk of not fully conveying the flavour of current archaeological thinking in these emerging interdisciplinary areas. Here the aim will be to make the backdrop as authoritative as possible so that the new results are set in the right context. Wherever there is obvious dispute over the implications of new results the links between the various disciplines will be recognised with an attempt to explain whether the varying interpretations of the data can be reconciled. So the analysis will emphasise how the new disciplines are altering our perceptions about prehistory, rather than fully exploring the ferment surrounding the wider archaeological debate.

1.4 WHERE DO WE START?

The question of the role of climate change in human evolution has been widely explored. This climatic influence extends back several million years into the Miocene era and involves many aspects of our links with other great apes and the progression of our species. Here the discussion is restricted to more recent times, and in particular the period covered since the emergence of anatomically modern humans (*Homo sapiens*, from now on referred to as 'modern humans'). As best we can tell the first modern humans appeared in Africa between 100 and 200 kya. Between 100 and 10 kya they spread into Eurasia, across to Australia and eventually into the Americas (Fig. 1.4). This period coincides largely with the time covered by the last ice age.

This restriction to more recent events will be more marked for most of our discussion. The arrival of modern humans in Europe around 40 kya coincides with the first widespread evidence of an important shift in human behaviour. Often termed the 'Upper Palaeolithic Revolution', the shift was reflected in more versatile stone blades and tools, wider use of other materials (antler, bone and ivory) for tools, and the emergence of figurative art and personal ornaments (Mellars, 1994, 2004; Bar-Yosef, 2002). While this step forward in human intellectual development may have occurred much earlier, as recent analysis of decorated objects from southern Africa suggests

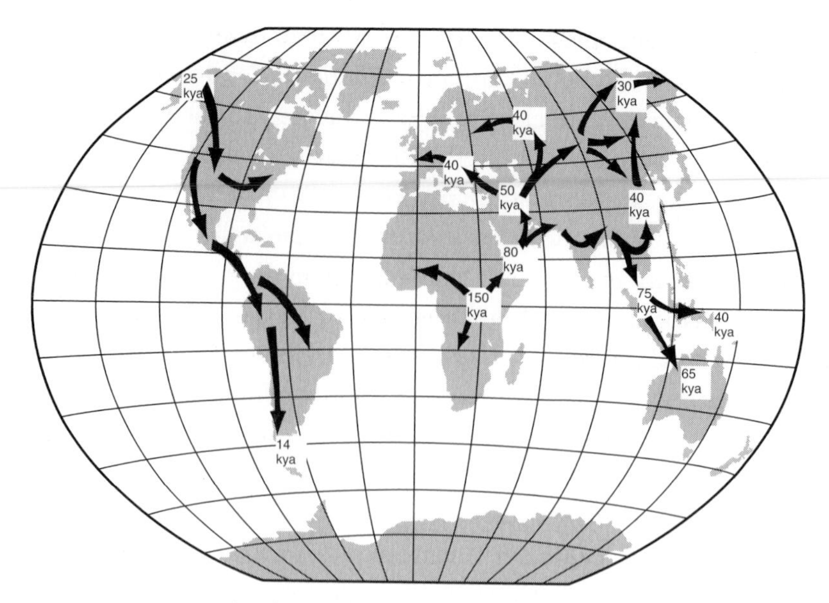

FIGURE 1.4 A schematic map showing the approximate timing of the migration of modern humans out of Africa and across the world in units of thousands of years ago (kya). (NB There is considerable disagreement about the timing of certain of these movements, notably in respect of the peopling of the Americas, which will be the subject of detailed analysis later in the book.)

(Henshilwood *et al.*, 2002), the sudden widespread emergence of this behaviour in the archaeological record provides a convenient starting point for much of what this book is about.

1.5 WHAT DO WE COVER?

The analysis here will inevitably place particular emphasis on the events in Eurasia and the North Atlantic. This is a consequence of two facts. First, in terms of human artefacts that reflect the emergence of modern humans and their intellectual development, the majority of evidence has, so far, been found in Eurasia. Second, the central role the Greenland ice cores will play in the analysis inevitably points the discussion in the direction of the North Atlantic and its surrounding landmasses. This geographical bias is, however, not a deliberate part of some Eurocentric model of human development, and wherever possible evidence of development in other parts of the world will be fully recognised to achieve as global a perspective as possible (Coukell, 2001).

In the context of climatic change, the emphasis on the North Atlantic does have a more reputable physical foundation. As will become clear, not only does the circulation of the North Atlantic carry so much more energy into the Arctic than does the Pacific, but also this circulation is capable of the dramatic and chaotic shifts that are central to this book. The fluctuations in this energy transport exert a fundamental control on the climate of the northern hemisphere. The extent to which events in the southern hemisphere are independent of this northern influence will be examined. In particular, how this affected the climate of much of Africa is an essential part of our story.

The importance of a global perspective is central to understanding the nature of the climate during the last ice age. It was not simply that it was so much colder then: parts of the globe experienced what we would regard as thoroughly acceptable climates even at the nadir of the last ice age, usually termed the last glacial maximum (LGM). No, it was that chaos reigned over the climate. So not only was much of northern Eurasia and North America buried under ice several kilometres thick, but the climate across the northern continents swung from the depths of glacial frigidity to relative mildness in the space of a few years. This erratic behaviour was a feature of virtually the whole of the last 100 kyr of the Earth's history. Then, after the final paroxysms of the ice age came to an end around 12 kya, the world warmed up dramatically over the next two millennia and settled down into a quiescent mode: in effect the reign of chaos ended.

The fact that the climate has been so relatively stable for the past 10 kyr (the *Holocene*) is widely recognised as the central reason for the explosive development of our social and economic structures. Once the climate had settled down into a form that is in many ways recognisable today, all the trappings of our subsequent development (agriculture, cities, trade etc.) were able to flourish. This change is often presented as the trigger for the onset of orderly progress and opportunities for 'great leaps forward'. When viewed in the longer perspective of our having evolved during the chaotic period prior to Holocene, this analysis may well be oversimplified, if not downright misleading.

The second and closely related aspect of our history is the growing evidence of the greater antiquity of many of the aspects of human intellectual development. These early developments will have been profoundly influenced by the climatic rigours of the time. So there are many ramifications for how social systems developed in the depths of the last ice age and the consequent implications for what we have now become.

The breadth of these issues mean that there has to be a self-denying ordinance on trying to cover all the fascinating issues in prehistory that could possibly relate to climate change. In this respect, restricting the story principally to events after the arrival of modern humans in western Europe makes good sense. To extend back into the subject of how this movement is related to the earlier development of archaic humans, notably Neanderthals, takes us into a different world. It is not just that this issue has been covered in considerable depth in a variety of previous books (Stringer & Gamble, 1993; Trinkhaus & Shipman, 1993). More importantly, recent genetic research strongly suggests that modern humans replaced archaic humans without interbreeding (Krings *et al.*, 1997; Krings *et al.*, 2000). So, since around 30 kya, as hominids, we have been on our own (Klein, 2003; Mellars, 2004). This means that since then, in terms of considering the impact of climate change on human prehistory, we are all that matters.

As far as the boundary of prehistory and 'recorded' history is concerned, the choice is bound to be arbitrary. The extent of records of both human activity and, crucially, climate-related events, varies from place to place. Furthermore, by definition 'Dark Ages', whether or not they were, in part, the product of climatic events, are less well recorded than the better times that preceded and followed them. There is, however, a sense of events being less well defined prior to the rise of the Greek and Roman Empires. In practice, there are virtually no documentary records of events that would illuminate the human consequences of climatic variability before them.

This does not mean that we are short of ways of measuring the climate of the time, as the earlier observations about the insights gained from ice cores demonstrate. What it does mean is that we are almost wholly dependent on 'proxy' records of the climate. Alongside ice cores these include tree rings, ocean sediments, coral growth rings and pollen records from lake sediments. Proxy data contain an amazing amount of climatic information, but they rarely if ever provide a direct measure of a single meteorological parameter. For instance, the width of tree-rings is a function of temperature and rainfall over the growing season, and also of groundwater levels reflecting rainfall in earlier seasons. Only where the trees are growing near their climatic limit can most of the growth be attributed to a single parameter (e.g. summer temperature). With other records (e.g. analysis of the pollen content, or the creatures deposited in ocean sediments), drawing climatic conclusions depends on knowing the sensitivity of plants and creatures to the climate and how their distribution is a measure of the climate at the time. Although a large proportion of the information of climate change in more recent times has been obtained from these forms of data, the change in the nature of documentary data around 2.5 kya allows us to make the arbitrary decision that in this book 'prehistory' effectively ends around this time.

In terms of population numbers, the period covered by the book involves perhaps the most interesting period of population growth of the human race. In the Upper Palaeolithic, around 35 to 30 kya, we numbered a mere few hundred thousand mortals. This number may have fluctuated dramatically through the LGM and then started to rise during the emergence from the ice age. By 10 kya the numbers had risen to around five million. The coming of agriculture and growth of civilisations led to a rise to around 100 to 150 million people by 2.5 kya (Kremer, 1993).[2]

[2] There is also useful information at the US Bureau of the Census, http://www.census.gov/ipc/www/worldpop.html.

1.6 CLIMATE RULES OUR LIVES

The hypothesis that climate change exerted a profound impact on the prehistoric development of human societies must be explored in terms of our current experience of how fluctuations from year to year have affected our own societies in more recent history. It is easy to forget how much our lives are defined by the climate of where we live and how vulnerable we are to extremes that fall well outside normal experience. Even in the seventeenth and eighteenth centuries there were major subsistence crises in Europe as a result of bad weather and poor harvests (Burroughs, 1997; pp. 34–39). The string of cold wet years in the 1690s brought disaster to farming communities across the continent. In Finland it is estimated that in 1697 the famine killed a third of the population.

Extreme weather has remained a dominant factor in the fortunes of agriculture right through the twentieth century. During the 1930s the drought across the Great Plains of North America (the 'Dust Bowl' years) caused immense social disruption. In our modern industrial world, we have done much to reduce this vulnerability. Nevertheless, we still have to design many features of our lives to handle whatever the climate throws at us over the years. So in understanding many aspects of societies around the world we need to know about the climate and our capacity to handle extremes. It affects everything from agriculture and the design of buildings through our industrial and transport systems to our diet and leisure activities. In the same way, any attempt to appreciate life in prehistory requires us to know as much as possible about the climate and its variability.

Consider the implications of evolving in a radically different type of climate. If, as seems to be the case, for more than 90 per cent of the time that our species has existed on this planet it has had to grapple with an immeasurably more capricious climate, the consequences for how we evolved are profound. Indeed, around 70 kya, we may have come perilously close to being wiped out by the hostile environmental conditions of the time. Our very ability to survive these challenges was a consequence of whatever skills we had then.

Furthermore, the combination of surviving these challenges and the process of natural selection must have ensured that the climate is deeply etched into our genetic make-up. It may also lurk deep within our psyche. As Ernest Shackleton observed, when contemplating being marooned on the ice of Antarctica in 1915, 'We had reached the naked soul of man.'

1.7 THE INTERACTION BETWEEN HISTORY AND CLIMATE CHANGE

Any discussion of how our growing knowledge of past climate change can influence our thinking about human prehistory must first consider what we mean by history. To most of us history is a matter of establishing what happened in the past. In practice, a great deal of historical analysis is more about how societies use knowledge of the past to inform thinking about current events. This transposition is potentially crucial to any discussion of prehistory and climate change. Much of our thinking about prehistoric human development is bound up with assumptions about progress and how we have evolved into much more advanced creatures than our palaeolithic forebears. Prehistoric art, and many other features of life long ago, require us to think more closely about how surviving the ice age can inform us about our behaviour now.

Our interpretation of the impact of climate change on prehistory is also bound up with current concerns about global warming (Intergovernmental Panel on Climate Change (IPCC), 2001a). Whatever messages can be extracted from the evidence of the past impact of climatic events on human development provides guidance on how to face up to the challenge of future climate change. So this particular form of historical analysis is every bit as much about using knowledge of the past to think about the present as it is about telling us what actually happened in the distant past.

2 The climate of the past 100 000 years

Here about the beach I wander'd, nourishing a youth sublime
With the fairy tales of science, and the long result of Time.

Alfred Lord Tennyson (1809–1892), *Locksley Hall*

The scale of prehistoric climate change is almost incomprehensible. The broad features of the last ice age are well known. For much of the past 100 000 years (100 kyr), tens of millions of cubic kilometres of ice were piled up on the northern continents. Where Chicago, Glasgow and Stockholm now stand there was ice over a kilometre thick. Sea levels were as much as 130 metres lower since huge amounts of water were locked up in the ice sheets, and vast areas of what is now continental shelf were exposed for tens of thousands of years. Then suddenly some 15 000 years ago (15 kya) things warmed up, briefly reaching levels comparable with more recent times, and the ice sheets began to disappear. Even then the climate took one last icy descent into the freezer. For several hundred years, conditions in the northern hemisphere suddenly plunged back to something close to the greatest glacial severity of the ice age. Finally, the climate truly changed for the better so that for about the past 10 kyr the Earth has experienced relatively benign conditions. Perhaps more than anything else, this change has enabled humankind to develop the social structures we see around us today.

The assumption of an ice-age climate, followed by much warmer conditions, is implicit in much of the thinking about the emergence of agriculture, the formation of settled communities and the subsequent development of ancient civilisations. While the broad outline is correct, it blurs out a vast amount of detail. Just how much detail has become fully apparent only in recent years, as a wide variety of research programmes have yielded much more information about the past.

What is now clear is that during the last ice age, and the period that followed it, the climate was much more chaotic than it has been in recent millennia. Generally, the climate was much more variable. Sudden changes occurred from time to time. Collapse of parts of the ice sheets, or release of meltwater lakes that built up behind the ice, led to cataclysmic changes. Armadas of icebergs or floods of icy freshwater swept out into the North Atlantic altering the circulation of the ocean at a stroke and with it the climate of the neighbouring continents. With a flick of the climatic switch, Europe and much of North America could be plunged back into icy conditions, having only just emerged from the abyss of the preceding millennia. Conversely, the stability of the glacial conditions could be interrupted by a re-establishment of the flow of warm water to higher latitudes in the North Atlantic, bringing surprising temporary warmth to the northern continents.

Just how rapid, and how large, some of these changes were is one of the great surprises of recent climate change research. Moreover, the implications for life around the world, and in Eurasia and North America in particular, are profound. Any analysis of prehistory that fails to take full account of these turbulent events may well miss essential aspects of human development. Yet the recent emergence of much of the evidence for these changes means that it will take time for this to register with other disciplines, such as archaeology and anthropology. In the meantime, interpretation of many of the arguments concerning human prehistory will be illuminated by knowing more about discoveries of climate change, how the measurements have been made, and how much confidence we can have in how the climate really changed in different parts of the world.

2.1 DEFINING CLIMATE CHANGE AND CLIMATIC VARIABILITY

Before we start on the exploration of climate change over the last 100 kyr it helps to have a set of definitions about what constitutes climate change and variability. Although this may appear to be dusty statistical hair-splitting, it is essential in unscrambling the extraordinary range of fluctuations that have occurred during the last ice age and the subsequent warm period. At the most basic level, we are not concerned about the weather, which is what is happening to the atmosphere at any given time, but about the climate – that is, what would be expected to occur at any given time of the year based on statistics built up over many years. Changes in the climate constitute shifts in meteorological conditions lasting a few years or longer. These changes may involve a single parameter, such as temperature or rainfall, but usually accompany more general shifts in weather patterns that might result in a shift to, say, colder, wetter, cloudier and windier conditions.

The next issue to address is defining the difference between climate variability and climate change. This may seem like an artificial distinction, but as we will see, it is important to spell out clearly how the two categories differ. Figure 2.1(a) presents a typical set of meteorological observations; this example is a series of annual average temperatures, but it could equally well be rainfall or some other meteorological variable for which regular measurements have been made over the years. This series shows that over the period of the measurements the average value remains effectively constant (the series is said to be *stationary*) but fluctuates considerably from observation to observation. This fluctuation about the average, or mean, is a measure of *climate variability*. In Fig. 2.1(b), (c) and (d) the same example of climatic variability is combined with examples of *climate change*. The combination of variability and a uniform cooling trend is shown in Figure 2.1(b), while in curve (c) the variability is combined with a sudden drop in temperature, which represents, during the period of observation, a once and for all change in the climate, and in curve (d) the variability is combined with a periodic change in the underlying climate.

The implication of the forms of change shown in Fig. 2.1 is that the level of variability remains constant while the climate changes. This need not be the case. Figure 2.2 presents the implications of variability changing as well. Curve (a) presents the combination of

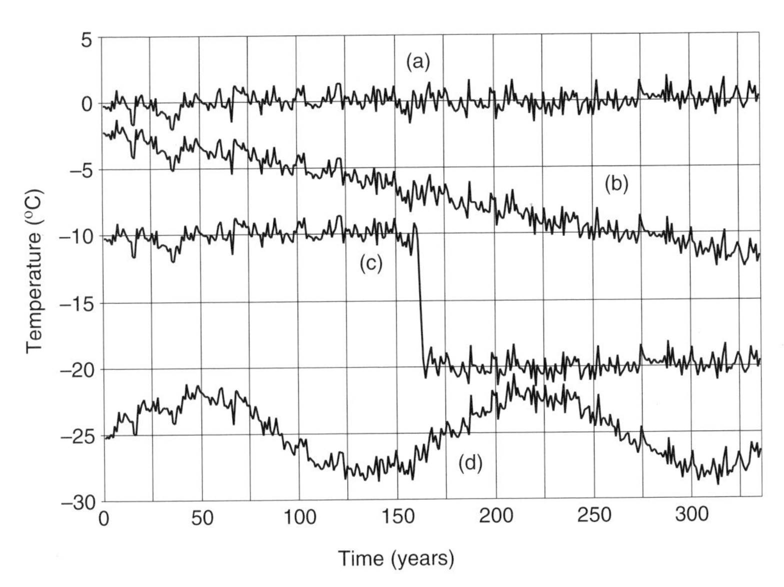

FIGURE 2.1 The definition of climate variability and climate change is most easily presented by considering a typical set of temperature observations which show (a) climate variability without any underlying change in the climate; (b) the combination of the same climate variability with a linear decline in temperature of 10 °C over the period of the observations; (c) the combination of climate variability with a sudden drop in temperature of 10 °C during the record, with the average temperature otherwise remaining constant before and after the shift; and (d) the combination of climate variability with a periodic variation in temperature of 6 °C. Successive records are displaced by an appropriate amount to enable a comparison to be made more easily.

the amplitude of variability doubling over the period of observation, while the climate remains constant. Although this is not a likely scenario, the possibility of the variability increasing as, say, the climate cools (Fig. 2.2(b)) is much more likely. Similarly, the marked increase in variability following a sudden drop in temperature (Fig. 2.2(c)) is a possible consequence of climate change. During and since the last ice age, in one way or another, all the forms of climate variability and climate change presented on Figs. 2.1 and 2.2 have occurred. They all represent examples of the challenges that the climate presented to human societies in prehistory.

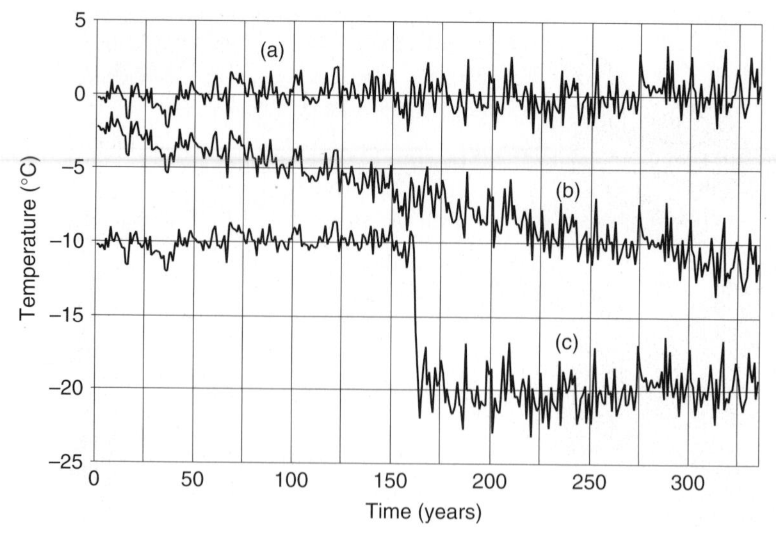

FIGURE 2.2. The combination of increasing climate variability and climate change can be presented by considering a set of temperature observations similar to those in Fig. 2.1 which show (a) climate variability doubling over the period of the record without any underlying change in the climate; (b) the combination of the same increasing climate variability with a linear decline in temperature of 10 °C over the period of the observations; and (c) one level of climate variability before a sudden drop in temperature of 10 °C, which then doubles after the drop while the average temperature remains constant before and after the shift. Each record is displaced by an appropriate amount to enable a comparison to be made more easily, and the example of periodic climate change in Fig. 2.1 (curve d) is not reproduced here as the nature of changing variability in such circumstances is likely to be more complicated.

2.2 THE EMERGING PICTURE OF CLIMATE CHANGE

How the current picture of the ice ages and their impact on the world around us was built up is a fascinating example of the tortuous progress of scientific thinking. The original proposals about the ice ages are associated with the Swiss naturalist Louis Agassiz. The possibility of ice sheets covering part of northern Europe had first been proposed by James Hutton, the founder of scientific geology, in 1795, and reiterated by a Swiss civil engineer, Ignaz Venetz, in 1821. In 1824 Jens Esmark, a Norwegian geologist, offered the theory that Norway's mountains had

FIGURE 2.3 Glacially polished rocks and morainic debris at the edge of the Zermatt glacier. From Louis Agassiz's *Études sur les glaciers* (Neuchâtel, 1840; *Cambridge Encyclopaedia of Earth Sciences*, Fig. 1.12.).

been covered by ice, and in 1832 a German professor of forestry, Bernhardi, published a paper suggesting, on the basis of the large number of erratics on the North German Plain, that a colossal ice sheet had extended from the North Pole to the Alps. But these ideas received scant attention.

In the summer of 1836, while on a field trip in the Jura Mountains with Jean de Charpentier, a friend of Ignaz Venetz, Agassiz became convinced that blocks of granite had been transported at least 100 km from the Alps (Fig. 2.3). In 1837 he first coined the term Ice Age (*die Eiszeit*) and in 1840 his proposals were published in a ground-breaking book. At first the geological community ridiculed the theory, but his passionate advocacy of the ice age was to prevail.

Agassiz travelled to Scotland where he saw more evidence of glaciation and then in 1846 arrived in Nova Scotia where again the evidence of ice was plain to see. In 1848 he joined the Harvard faculty and was active in many fields, notably marine science, but continued glacial research in New England and around the Great Lakes. Over the

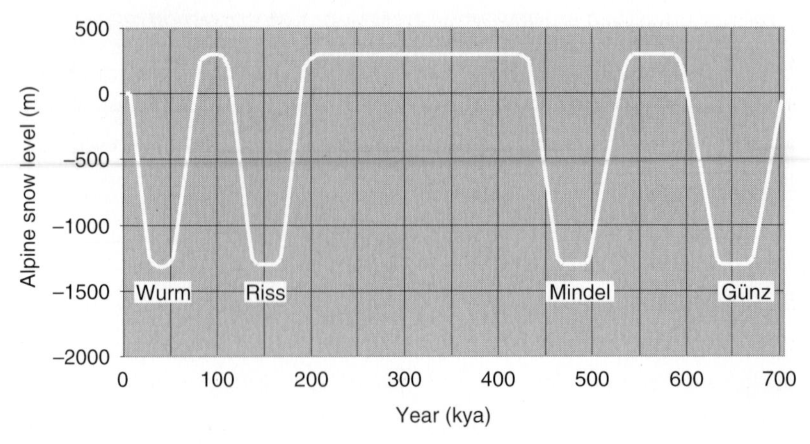

FIGURE 2.4 Penck and Brucker model of past ice ages, which was based on studies of the geological evidence of fluctuations in the snow level in the European Alps and was published at the beginning of the twentieth century. The four main ice ages (Wurm, Riss, Mindel and Günz) were named after places in the Alps that showed clear evidence of each particular glacial period. In other parts of the world these ice ages are known by place names that were identified as good examples of this succession of events.

next few decades a variety of geological evidence made it clear that many features of the northern hemisphere could only be explained by ice ages and Agassiz was vindicated, although in some quarters the subject remained controversial until the end of the nineteenth century.

Following the work of Agassiz an orderly view built up during the remainder of the nineteenth century about ice ages. This was that over the last six hundred thousand to a million years there had been four glacial periods lasting around 50 kyr separated by warm interglacials ranging from 50 to 275 kyr in length (Fig. 2.4). The present interglacial started about 25 kya and was destined to last at least as long as previous interglacials. The principal evidence of the orderly progression was seen in glaciated landscapes of the northern hemisphere. The scoured U-shaped valleys, eroded mountains, drumlins (mounds of stiff clay moulded under, and by, the creeping ice), eskers (ridges of gravel and sand formed by the meltwater flowing out of edges of the ice sheets), large glacial boulders (erratics), glacial tills and terminal moraines (material bulldozed by the front edge of glaciers) are obvious landscape features. While some other geologists were more cautious about the chronology of the events, the broad picture of four major glaciations was the accepted interpretation of the geological evidence in both Europe and North America.

This simplified view of the past underpins a set of popular images of last ice age. In these icy periods, which gripped the world for tens of thousands of years, mammoths were seen trudging through a never-ending snowscape, across which blizzards raged, while in the background the ominous edge of a permanent ice sheet lurked. As for our predecessors, any of them who were tough enough to survive in these frigid wastes lived in caves, hunted the huge beasts that roamed these regions, and wore an assortment of animal skins in a vain attempt to keep out the cold. There had been occasional warmer periods. The possibility of rapid and frequent changes from intense glacial cold to relative warmth seems not, however, to have entered this orderly version of events. Furthermore, even during the coldest periods, the long bitter winters were punctuated by short relatively warm summers, when abundant grass, hardy shrubs and dwarf trees grew on the steppe landscape: mammoths and woolly rhinoceros needed a lot of fodder to keep them going. In the same way as northern North America and Siberia supported extensive megafauna throughout the Holocene, and Africa does now, during much of the last ice age vast areas of mid-latitude Eurasia sustained abundant wildlife.

This stereotypical and blinkered view of the past reflected the limitations of the data. On land, many features of the past are swept away by subsequent events. This is especially true of anything to do with ice sheets. As they waxed and waned they scoured the landscape clean of soil and the attendant vegetation. Around their margins, evidence of changing weather and shifting vegetation types was either removed for some periods, or during the coldest intervals there was no identifiable build-up of deposits. So, much of the evidence was obliterated and forming an accurate local picture of the sequence of past events is difficult or impossible. Nevertheless, inklings of shorter-term climatic changes were evident in early studies. The evidence of the dramatic changes at the end of the ice age was clear in pollen records extracted from lake sediments in Denmark: the rapid cooling noted in the introduction to this chapter is known as the Younger Dryas, after a small arctic plant that was a clear indicator of arctic tundra conditions. Similarly, milder periods during the ice age, known as *interstadials*, were identified across Eurasia and North America and given local names, such as Briansk, Denekamp, Hengelo and mid-Wisconsin. But it was not possible to provide a coherent global picture of how these events related to one another and whether they represented the sum total of the most significant variations in the climate during the last ice age.

2.3 PROXY DATA

Our view of climate history started to change radically in the 1950s. This was built on a variety of sources, such as tree rings, pollen records, ocean sediments and ice cores. Known as *proxy data*, measurements of the properties of these sources provided new insights into climate change. Although tree rings and pollen records had already provided insights into past climate change, it was the advent of new technologies, such as the ability to drill cores from the sediments at the bottom of the deepest ocean basins, that transformed our knowledge of the world around us. One of the first climatic products of this type of work was a set of papers by Cesari Emiliani, at the University of Chicago, on the properties of fossil shells of the tiny creatures found in the sediments of the tropical Atlantic and Caribbean (Emiliani, 1955). Using the reversal of the Earth's magnetic field around 750 kya as a marker he was able to show that, instead of just four ice ages punctuated by lengthy interglacials, there had been seven glacial periods since then, occurring every 100 kyr or so.

A period of intense debate followed about the validity of the ocean sediment data. Gradually, however, a growing body of evidence from ocean sediments around the world, together with pollen records from part of Europe that had not been covered by ice, and ice cores

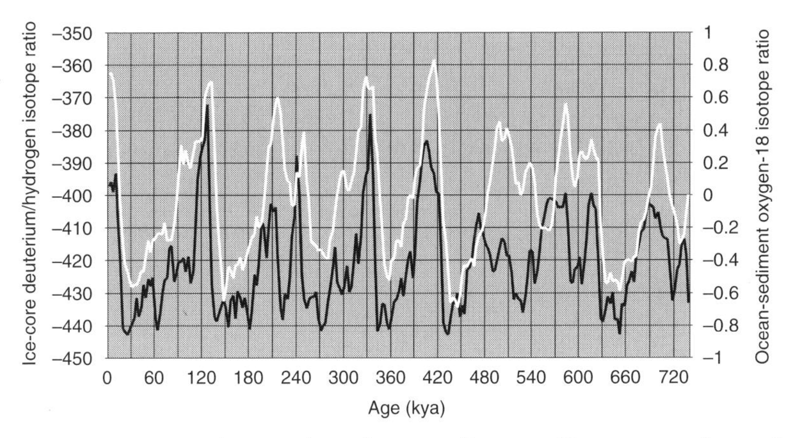

FIGURE 2.5 A comparison of ocean-sediment and ice-core records over the past 736 kyr (an increase in isotope ratios means an increase in temperature). The white line shows the proportion of oxygen-18 ($\delta^{18}O$) per thousand (‰) in the foraminifera sampled in 13 ocean sediment records (from Karner *et al.*, 2002) and the black line shows the change in the deuterium/hydrogen isotope ratio measured in the EPICA ice core drilled at Dome C in Antarctica (data from EPICA Community Members (2004), supplementary information, www.nature.com/nature).

from Greenland and Antarctica confirmed Emiliani's conclusions. The dramatic message from the deep-sea cores was that beyond any shadow of reasonable doubt the climate of the last million years has been dominated by periodic variations in the Earth's orbital parameters that have resulted in ice ages occurring roughly every 100 kyr. Subsequent work has led to the stacking of many sediment records from the open ocean to produce a standard curve for changes in oxygen isotopes for well over the past 740 kyr (Fig. 2.5; Karner *et al.*, 2002).

There are two principal sources of climatic information in the cores. The first is the ratio of oxygen isotopes in the calcium carbonate in the skeletons of the foraminifera living in the deep water. Changes in this ratio are a longer-term consequence of how the isotope content of precipitation locked into the ice caps of Greenland and Antarctica (see Section 1.2) varies with temperature. As the ice sheets in the northern hemisphere grew, the amount of ^{16}O locked in the ice was proportionately greater than the amount of ^{18}O. So the ratio of these oxygen isotopes in the oceans changed as the amount of ice rose and fell, and these variations were reflected in the shells of the foraminifera, which formed using oxygen in the oceans. This means that changes in the $^{16}O/^{18}O$ ratio are a measure of the size of the ice sheets at high latitudes, and hence indirectly of fluctuations in global temperatures.

The second important source of climatic information is the types of species living near the surface at any given time, and their relative abundance is a guide to surface temperatures – the presence or absence of species that only survive in warm or cold water is a direct measure of the conditions when they were alive. So by mapping the populations of different creatures at different times using cores taken from around the world, it is possible to construct a picture of how the temperature of the oceans' surface waters varied over time. In addition, other climatic information can be extracted from these sediments. Notably, the presence of larger bits of rock provides information about how far glacial debris has been transported out into the oceans by icebergs, while changes in the size of grains of sand can provide insights into the strength of currents.

A key element in the presentation is a set of agreed stages in the chronology of the oxygen isotope record in the ocean sediments. These 19 stages are used to define the principal glacial and interglacial periods since the Matuyama–Brunhes magnetic reversal around 750 Kya. The last interglacial and the subsequent ice age, together with the warming since the LGM, are covered by five defined stages designated Oxygen Isotope Stages One to Five (OIS1 to 5). This is a useful standard nomenclature to adopt when considering the conditions on the northern continents during the last 130 kyr.

Ice-core measurements described briefly in Chapter 1, which have come from ice sheets around the world, are another invaluable source of information. Just how much these ice-core records have altered our perceptions is evident from the latest core drilled by a European team (EPICA), high on the ice sheet of Antarctica. This core provides a continuous record for over 730 kyr and covers the last eight ice ages (EPICA Community Members, 2004). It shows that cold conditions prevailed for much of the time, especially during the last

400 kyr, and were interspersed by much shorter warm interglacials (Fig. 2.5, black curve). This pattern was clearly a global phenomenon as the same features are evident in the results from ocean-sediment cores in the tropics (Fig. 2.5, white curve).

Since around 400 kya the pattern has been broadly consistent. Each ice age ended rapidly, then after an interglacial that lasted only around 10 kyr, the climate slipped back into glacial conditions, giving the temperature record a characteristic sawtooth appearance. In this record we see examples of three forms of climate change identified in Fig. 2.1: sudden changes in the form of rapid rises in temperature, long slow periods of cooling and an overall periodic fluctuation in the climate of the interglacial periods occurring every 100 kyr. Superimposed on this broad pattern are shorter-term periodicities that occur about every 20 kyr.

The records in the northern hemisphere obtained from both ocean sediment and ice-core data for the period covering the last glacial provide a more detailed picture. The striking suddenness of many of the changes came as a surprise to many climatologists. The way in which the conditions shifted almost instantaneously between radically different conditions was dubbed (Taylor *et al.*, 1993) a 'flickering switch'. Equally impressive were the dramatic fluctuations in temperature on millennial timescales. All these can be seen in the Greenland ice core (Fig. 2.6).

There is, however, one further underlying issue to address. This is the question of the reliability of the clocks used to measure the timing of events in the past. We have already talked about ocean sediments and ice cores, both of which rely on making informed guesses of the rate at which these records built up. When we extend our climatic studies to include other proxy records, such as tree rings, corals, lake sediments, and stalactites and stalagmites, in the best possible conditions we can make precise estimates of the rate of growth, as in the case of the annual growth rings in trees. This does require some clever detective work to build up series including long-dead trees to extend the series back over thousands of years, but it is possible in some parts of the world.

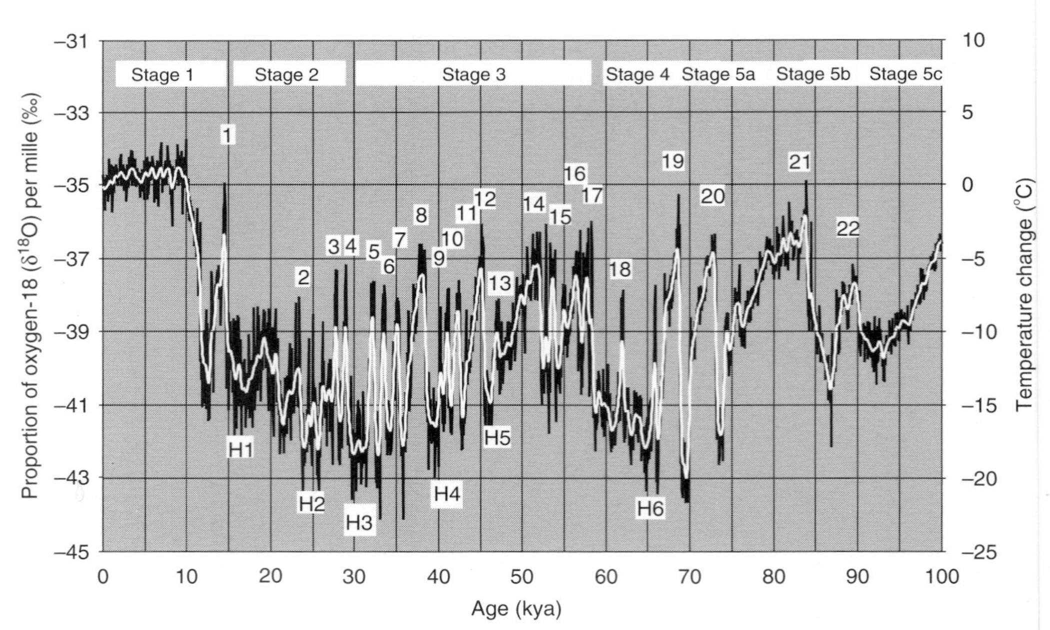

FIGURE 2.6 The record of the proportion of oxygen-18 ($\delta^{18}O$) per thousand (‰) in the GISP2 ice core (the black line is the data average values for every 50 years and the white line is the 41-term binomial smoothing of these data). These curves show the 20 Dansgaard/Oeschger warming events (labelled 1 to 20), six of which combined with Heinrich events (labelled H1 to H6). The temperature range covered by these changes is reckoned to be about 20 °C between the coldest periods and the warmth of the last 10 kyr. The various Oxygen Isotope Stages identified in the ocean-sediment core analysis are shown at the top of the diagram. (Data archived at the World Data Center for Paleoclimatology, Boulder, Colorado, USA.)

Similarly, ice cores contain distinctive annual layers, provided there is no appreciable melting in the summer. But as we go deeper and deeper into the ice sheet the layers get thinner and thinner as they are extruded out sideways by the weight of the ice above them. So at some point it becomes increasingly hard to count individual layers with any reliability and we have to turn to models of how the ice will spread out over time (see Appendix). When we are measuring the layers of sediment deposited on the bottoms of the oceans or lakes then we have to make assumptions about whether the rate of deposition is constant and whether burrowing creatures have disturbed the layers.

The challenges are not insurmountable as there are other means of dating that come to the palaeoclimatologist's rescue. These include radiocarbon dating and the use of other naturally occurring radioactive isotopes to provide an absolute check on the dating processes (see Appendix), plus arcane techniques such as thermoluminescence and electron spin resonance (ESR). The physical details of these techniques are not of direct concern here. What matters is that, while this plethora of measurement techniques has built up a formidable armoury of methods of dating, they often produce different answers for the same events. As a consequence, at the minimum there is some uncertainty about the precise dates of events whilst in some cases there may be considerable doubt about whether we are even talking about the same event. From now on the reader is advised to pay close attention to hints of uncertainty about the correspondence of events as it may reflect a more fundamental weakness in the analysis.

2.4 DO ICE-CORE AND OCEAN-SEDIMENT DATA RELATE TO HUMAN EXPERIENCE?

One fundamental aspect of analysing climate change must be addressed from the outset. This is whether the evidence from far-flung places like Antarctica, Greenland or the depths of the great oceans provides a real measure of what was happening across the plains of Eurasia, or in Africa or Australia. Clearly the major changes associated with the waxing and waning of the ice ages were global in nature.

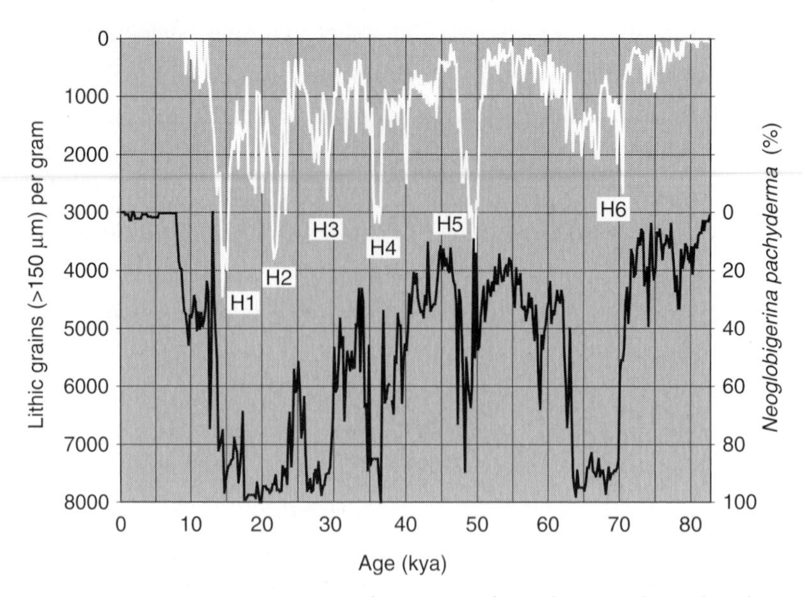

FIGURE 2.7. Deep-ocean sediment core from the central North Atlantic showing the link between Heinrich events (labelled H1 to H6), which can be seen in the increase of ice-rafted debris (lithics) (white curve) and the surface water temperature as revealed in the incidence of cold-water foraminifera *Neoglobigerina pachyderma* (black curve). (Data archived at the World Data Center for Paleoclimatology, Boulder, Colorado, USA.)

The real question is, however, how the more rapid changes affected different parts of the world. Only if we can be sure that the most important features of climate change including its short-term variability were truly part of life across the continents of the world can we be confident of using these invaluable sources to underpin our analysis.

The first stage in this comparison is to look at the differences between the Greenland ice-core data and the results obtained from ocean sediments in the North Atlantic. A typical set of results for this region is shown in Fig. 2.7 for a site roughly midway between Newfoundland and southern Ireland.[1] The first thing to note is that there is much less detail than in the ice cores, as the ocean sediments have at best a resolution of several hundred years. This is because not

[1] The core used for this figure is designated DSDP-609 and the data are archived at the World Data Center for Paleoclimatology, Boulder, Colorado, USA.

only is the rate of deposition of these sediments much slower than the build up of ice cores (decadal accumulation rates typically being of the order of a millimetre rather than a metre), but also burrowing creatures searching for food stir up the top layers of the ooze. Only where conditions near the seabed are anoxic is the deposition undisturbed (see below). Nevertheless, many of the major features are the same, and the differences are, in part, a reflection of the difference in what the data are recording. In the case of the cold-water foraminifera (*Neoglobigerina pachyderma*) the variation from 0 to 100% covers the range from warm surface waters, typical of the Holocene, to the ice-filled waters of the severest stages of the ice age. So this curve picks out the coldest periods most clearly. The truly notable feature is the additional emphasis given to the prolonged cold period from 70 to 63 kya, to the cold spell at around 36 kya and to the frequency of extreme cold from 30 to 15 kya. The top curve shows the amount of ice-rafted debris found in the sediment, which highlights the times when armadas of icebergs floated out into the North Atlantic (see Section 2.5).

The next point to consider is that where the ocean cores are taken close enough to land they provide direct insight into regional weather patterns that may extend well inland. This applies in the case of the North Atlantic, where the prevailing westerly circulation means that inferences about climate change from ocean cores can be extended across Europe and into central Asia. Equally well, cores in the northwestern Indian Ocean show laminated organic-rich bands, reflecting strong monsoon-induced biological productivity (Leuschner & Sirocko, 2000). In contrast, periods of lowered southwest monsoonal activity produce bands poor in organic carbon. This provides valuable insights into the nature of the monsoon circulation over much of southern Asia. Furthermore, there is clear evidence of a close correlation between events in Greenland and the North Atlantic from the variations in the laminations in the cores in the northwestern Indian Ocean. This supports the conclusion that the fluctuations observed at high latitudes were also a significant component of low-latitude climate change during the last ice age. Moreover, these millennial and centennial fluctuations have been reflected in the strength of the monsoonal circulation over the Indian subcontinent.

Another prolific source of information is from the Cariaco Basin, off the coast of Venezuela (Petersen *et al.*, 2000). Here rapidly deposited (~3 mm per decade) organic-rich sediments contain visible annual laminations. What is more they are devoid of preserved seabed (benthic) faunas. This means the sediments were laid down in anoxic conditions, so there were no burrowing creatures scrambling the evidence. These deposits are composed of light (plankton-rich) and dark (mineral-rich) layers. This provides insights into what was going on in the surface waters as a result of strong seasonal fluctuations in trade-wind-induced upwelling and regional precipitation. The mineral sediments come from the surrounding watersheds and provide an accurate picture of regional hydrologic conditions. Cores from the Cariaco Basin have been used to provide detailed analysis of interannual climate change over the last 15 kyr, and of decadal fluctuations over the last 500 kyr.

Turning to land-based measurements, there is a huge range of independent observations that can be used to draw parallels with ice-core and ocean-sediment data. One of the most relevant sources of detailed climatic analysis has been obtained from lakes where layers of silt have been deposited for over 100 kyr. In these instances, and in many other cases of measurements on land, some of the most important indicators of changing environmental conditions are the amount and types of pollen found in sediments. If these sediments have been laid down in a regular manner, the abundance of pollen from different species of trees and shrubs provides details of climate change. Because different species have distinct climatic ranges, it is possible to interpret their relative abundance in terms of shifts in the local climate. So pollen records from many parts of the world have the potential to fill the geographical gaps that ice cores and ocean-sediment records cannot cover.

Pollen records are also of great historical importance in studying climatic change because of the early work charting the emergence of northerly latitudes from the last glaciation. Most pollens (from flowering trees and plants) and spores (principally ferns and mosses) are tiny. Few exceed 100 μm (0.1 mm) in diameter and the majority are around 30 μm. They have a waxy coat that protects them from decay. The size and shape of this outer wall, along with the number and distribution of apertures in it, are specific to different species and can be readily identified under a microscope.

Early work on pollen records focused on alterations in regional vegetation since the end of the last ice age. This work has established that these shifts were controlled by broad global patterns of climate change. For changes extending back into the last ice age there was until recently greater uncertainty as to whether other factors involving the migration and competition of species might have been more significant. There are two principal explanations for this uncertainty. First, almost all the longer pollen records related to northern Europe, so there were doubts about whether the results were representative of global changes. Second, many of the cores contained high-frequency variations during the last glaciation. These raised doubts about the dating of the strata in the cores, some of which had substantial gaps. It was not until more recent ice-core and ocean-sediment data became available that it was possible to check the timescales of these different records.

Recent pollen records obtained from cores drilled as far apart as the Massif Central in France (Thouveny *et al.*, 1994), southern Italy (Allen *et al.*, 1999) and Carp Lake in the Cascade Range in northwest USA (Whitlock & Bartlein, 1997) have enabled comparisons to be made across the northern hemisphere. These have confirmed that pollen records can be accurately dated back to the last interglacial, some 125 kya, and provide an extraordinary amount of additional information about how the climate has changed. The correlation between these results is remarkable. So, variations observed in pollen records can be combined with data from ice cores and ocean sediments to form a more detailed picture of global patterns of climate change during the last ice age and since then. In addition, measurements of the magnetic susceptibility of the cores in France and Italy also provide confirmation of the climate changes observed in the pollen records. This is because under cold climate conditions the freeze–thaw alternations caused the erosion of local rocks and the deposition of clastic sediments. Under temperate climatic conditions vegetation and soil development enhanced the organic content of the sediment, thereby diluting the magnetic fraction of the sediments. So magnetic susceptibility is strongly sensitive to local climate and provides an independent check on the conditions.

Another increasingly fruitful source of climatic data is beetle assemblages (Coope *et al.*, 1998). Beetles form roughly three-quarters of all animals found in terrestrial and freshwater-brackish environments. Their habitat is dependent on climatic factors, rather than being plant-specific, and when the climate changes they swiftly migrate to more congenial localities. Furthermore, their remains are well preserved in sediments. So drawing on present-day data of the distribution of beetle species it is possible to reconstruct the probable temperature regime represented by a fossil sample. Using information on several species can often refine this analysis. The temperature information usually combines the mean summer maximum and the range between summer and winter to draw detailed inferences about past climates on the basis of the distribution of different beetle species in the fossil record.

Rather different climatic information is recorded in the formation of stalactites and stalagmites. The deposition of calcium carbonate by running water in caves forms encrustations (*speleothems*) that can accumulate at approximately constant rate. Similar to the way in which the properties of foraminifera are affected by the isotope ratio of the water they grow in, speleothems are affected by the changes in the isotopic ratios in precipitation. Where such encrustations build up over a long time, they have the potential to provide useful climatic information. They also have the advantage that, because rainwater dissolves uranium and its daughter products, speleothems can be dated absolutely by measuring their uranium/thorium ratio (see

Appendix). Their disadvantage is that where the amount of rainfall falls below a critical level, or in the case of the depth of the ice age turns to snow and becomes locked in ice and permafrost, the encrustation ceases. So breaks in the record require careful calibration, which is possible by measuring the uranium/thorium ratio throughout the speleothem. This provides an absolute dating of when different layers were deposited. In recent years an increasing number of measurements of speleothems from across the continents of the world have been published. These provide valuable confirmation of the coherent picture of the detailed fluctuations in the climate throughout the last ice age.

The breadth of these measurements provides the adequate opportunity to crosscheck whether the various aspects of climate change observed in the Greenland ice cores and the northern Atlantic sediment records do indeed extend to the rest of the northern hemisphere. Furthermore, the recent pollen and speleothem records have sufficient resolution to identify rapid changes in the climate that can be correlated with the sudden shifts in Greenland and the North Atlantic. This combination allows us to build up a comprehensive picture of the fluctuations that were so much a part of the ice age.

2.5 CHANGES DURING THE LAST ICE AGE

Armed with confidence in the utility of the Greenland ice-core data we can get into more detail about climate across the northern hemisphere during the last ice age. In particular the isotopic temperature records (see Fig. 2.6) show some 20 interstadials, which have now become known as 'Dansgaard/Oeschger (DO) events' after the scientists who first identified them (Dansgaard & Oeschger, 1989): the precise number is subject to slight variations from analysis to analysis depending on whether or not different climatological groups award the accolade of being an 'event' to the most transient warmings between 15 and 100 kya. Typically the events start with an abrupt warming of Greenland of some 5 to 10 °C over a few decades or less. This warming is followed by a gradual cooling over several hundred years, and occasionally much longer. This cooling phase often ends with an abrupt final reduction of temperature back to cold ('stadial') conditions. The overriding feature of these observations is that they constitute a much more complicated picture of the nature and frequency of interstadials than was realised from the earlier archaeological records.

Opposite in sign to the interstadials are extreme, relatively short-lived particularly cold events, known as 'Heinrich events', which were first recognized as periods with substantial ice-rafting in the North Atlantic (see the lithics curve in Fig 2.7; Heinrich, 1988; Bond *et al.*, 1992; Bond & Lotti, 1995). These events occur during stadials, and represent the brief expression of the most extreme glacial conditions. They show up in the Greenland ice cores as a further 3 to 6 °C drop in temperature from the already cold glacial climate, and these events coincide with particularly cold and arid intervals in European and North American pollen records. There is also evidence further afield of these events exerting a global influence on the climate, although the greatest impact was felt around the North Atlantic and in mid-latitudes oceanic surface productivity dropped precipitously.

These specific fluctuations are, however, only part of the story. As the ice sheets grew, and then fluctuated with changing climatic signals, their extent would have varied appreciably from place to place and time to time around the northern hemisphere. This means the scale and timing of climate changes varied from place to place. So, while the Greenland ice-core fluctuations are an indicator of how the climate in mid-latitudes of the northern hemisphere varied, in later chapters, when it comes to illuminating living conditions in any particular part of the world, any additional regional or local information will be grist to the mill, set in the context of the ice-core background.

Broadly speaking the principal features of the various records match pretty well during the early stages of the ice age (OIS 5c to 5a in Fig. 2.7). Thereafter, the increasingly erratic nature of climate change, as seen in the Greenland ice cores, has until recently been less easy to discern in the land-based data. In Europe, north of the Alps (e.g. eastern France), the pollen data for nearly all of OIS4 (74 to 59 kya) is usually presented as being one of sustained periglacial conditions. For much of the period, the record from the North Atlantic suggests that the region did suffer sustained cold after Heinrich event 6 (see Fig. 2.7). What is missing from this general picture is the dramatic swings in the ice-core records associated with the two striking DO warm events (numbers 19 and 20 in Fig. 2.6). There is some evidence of these extreme climatic fluctuations in recent pollen records from Lago Grande di Monticchio in southern Italy (Allen *et al.*, 1999) and a speleothem record in Hulu Cave in China (Wang *et al.*, 2001). A recent speleothem study in southwest France, however, shows relatively little evidence of these short-term events and a sustained cold period from around 67 to 61 kya (Genty *et al.*, 2003), which coincides closely with the figures obtained from the GISP2 record The implications of shorter-term differences in the various records will be explored when discussing regional climates in Chapters 3 and 4.

The various warmer periods during the OIS3 period (59 to 28 kya) are broadly reflected in the land-based data. The five interstadials that are usually attributed to this period can, with a little pushing and shoving of the timescales, be lined up with the main periods of warming in the Greenland climate.[2] But the fact that this period includes 15 DO events, which are not easily shoehorned into five interstadials, is a measure of the challenges thrown up by the ice-core data. In spite of the fact that reconciling the details of the various records will take a lot more research, the land-based data are central to our understanding of conditions across Eurasia during the ice age. For instance, during the Hengelo interstadial, between 38 and 36 kya, the climate across the northern Europe improved to the extent that deciduous forests spread into southern France and the Hungarian plain, while open coniferous forest grew as far north as the 50th parallel from Brittany to southern Russia.

[2] In Europe these interstadials are usually defined as Oerel (58–54 kya), Glinde (51–48 kya), Moershoofd (46–44 kya), Hengelo (39–36 kya) and Denekamp (30–28 kya).

In North America the available data provide a less detailed picture of conditions across the continent during the Stage 3 period. In part, this may reflect the dominant part played by the Laurentide ice sheet which affects the interpretation of data from the mid-west United States. In particular, there is little evidence of the impact of the various DO oscillations, and the principal warming appears to have occurred around 43 to 39 kya, an interstadial that in North America is usually known as the 'mid-Wisconsin'. It is, however, a measure of the relative lack of detail in the North American chronology that this term is often used to cover the entire period of Stage 3.

The striking summer warmth during some of the interstadials brings out an important feature of these warmer intervals. This is that, across most of the mid-latitudes of the northern continents, much of the drop in annual temperature occurred in the winter half of the year. These differences, in part, reflect the fact that for much of the ice age the northern ice sheets were not as extensive as during its final stages. They were, however, compounded by the second more basic fact: the Sun still stood high in the sky every summer and so the northern continents received abundant heat in summer. Far from the chilly influence of the icy northern oceans this warmth would have had a dominant impact on summer weather. So, although the summers may have been shorter, where the winter snow melted they would be relatively warm, just as is the case in Siberia nowadays.

The uncertainties about the nature of the climate in different parts of the world during OIS3 do not extend to the final stage of the ice age (OIS2). All the available records confirm that the glacial conditions reached their nadir between 25 and 18 kya: the Last Glacial Maximum (LGM). Although this was the coldest period of the ice age both the Greenland ice-core and North Atlantic sediment records suggest that the millennial-timescale fluctuations largely died away. Instead, it was a period of unrelenting cold, although short-term fluctuations on the decadal timescale remained substantial. The ice sheets reached their greatest extent around 21 kya. At this time, ice sheets up to 3 km thick covered most of North America, as far south as the Great Lakes, all of Scandinavia and extending to the northern half of the British Isles and the Urals. In the southern hemisphere much of Argentina, Chile and New Zealand were under ice, as were the Snowy Mountains of Australia and the Drakensbergs in South Africa. The total amount of ice locked up in these ice sheets has been estimated to be between 84 and 98 million cubic kilometres as compared to the current figure of about 30 million cubic kilometres. This was sufficient to reduce the average global sea level by about 130 metres (see Section 2.10).

The global average temperature at the height of the last ice age was at least 5 °C lower than current values. Over the ice sheets of the northern hemisphere the cooling was around 12 to 14 °C. The position in the tropics is less clear. A major international study (CLIMAP) in the 1970s concluded that temperatures were much closer to current figures and over some of the tropical oceans there might even have been warmer conditions (CLIMAP Project Members, 1976; CLIMAP, 1981). The fact that the magnitude and even direction of temperature change was different between different latitudes came as a surprise. The largest inferred changes were in the mid-to-high latitudes, especially in the North Atlantic region bordered by large ice sheets. Sea ice covered much of the Greenland and Norwegian seas and persisted through the summer. Seasonal meltback of sea ice was also reduced in the Southern Ocean. Upwelling systems associated with eastern boundary currents, such as off northwest Africa, led to significant cooling.

The real surprise was in the limited cooling near the Equator in all oceans. Most of all, the western Pacific temperatures appear to have stayed essentially the same as modern conditions and in some cases were even warmer. Overall, the inference was that the Earth cooled surprisingly little during the ice age, except at high latitudes. This concept of 'polar amplification' of climate change implies a feedback mechanism driven by increased albedo associated with expanded snow and ice cover.

On land, forests shrank and deserts expanded, and the ice sheets reached their maximum extent. The question of just how thick the northern ice sheets were has been the subject of much discussion with opinions tending to be divided into two camps (the 'maximalists' and the 'minimalists'). It is sufficient to say here that given how much the sea level fell in the LGM (see Section 2.10), either way the ice sheets exerted a massive influence on the climate at high latitudes.

During the last two decades the view of the conditions in the LGM has developed.[3] The growing body of knowledge emphasises the complexity of the changes taking place during the LGM, which is now defined to have occurred between 23 and 18 kya. There are two principal reasons for its being hard to reach firm conclusions about the nature and timing of the changes during the LGM. First, there is a basic measurement challenge of reconciling the results of different measurement techniques applied to a wide variety of proxy records. Second, there is no doubt that the lowest temperatures occurred at different times around the northern hemisphere. The rapid advance in the development of computer models of the climate has, however, made it much easier to explore the nature of the climatic conditions at time. Using the range of proxy measurements that are available it is probable that these models will enable us to find out much more about the conditions during the LGM.

The important shifts in our picture of the LGM also relate to the tropics. Although the general cooling of the oceans appears to have been no more than a degree or two, the tropical lowlands cooled by between 2.5 and 3 °C. More important, at higher altitudes the cooling may have been about 6 °C. This shift would have resulted in a weaker tropical hydrological cycle, which is consistent with the evidence of desiccation in many parts of the tropics. For the rest, both the seasonal and the regional variations around the world are best considered in the context of life in the ice age in the next chapter.

[3] Mix, Bard and Schneider (2001) present a review of the work of EPILOG (Environmental processes of the ice age: land, oceans and glaciers), which was set up in 1999 to update the analysis of CLIMAP.

2.6 THE END OF THE LAST ICE AGE

Any discussion of the worldwide changes associated with the end of the ice age has to come to grips with the rather quaint nomenclature used to describe these events. This nomenclature reflects the fact that botanists did the first work in this area. The widespread use of peat as fuel in Scandinavia in the nineteenth century exposed tree stumps and layering in the peat bogs. Clearly these deposits were the product of major climatic changes in the past. During the early part of the twentieth century, botanists in northern Europe and North America built up an extensive array of information from the vegetation, insects and pollen records from both peat beds and the sediments of lakebeds. This monumental effort produced detailed but somewhat separate chronologies for events. Here we will concentrate on the Greenland ice-core data to discuss the basic features of the transition between the LGM and the Holocene.

The arcane language evolved around the localities where the most important climatic horizons were identified. So the lexicon resonates with a strange mixture of the archaeological geography of the northern hemisphere and botanical terminology. The Norwegian biologist, Alex Blytt, made the first studies in the 1870s. The most frequently used terminology is, however, the product of the work of the Swedish botanist, Lennart van Post, who studied the peat deposits in glacial lakes, and in 1916 established a set of pollen 'zones' representing the botanical history of northwestern Europe since the end of the last ice age. He named the different climatic periods in terms of distinctive species of plants in the climate record. So, for example, the term *Dryas* denotes colder periods when this characteristic plant of tundra-like environments of the Arctic moved southwards after times of relative warmth.

The first act in this unfolding drama at the end of the LGM was a collapse in part of the Laurentide ice sheet. This led to a surge of icebergs out into the North Atlantic and the last Heinrich event around 16.5 kya (see Figs. 2.6 and 2.7). Pollen records for northern Europe show that this last cold interval was followed by a sudden and profound warming (known as the *Bølling*) around 14.5 kya, which tallies with the rise in temperature seen in Fig. 2.6.

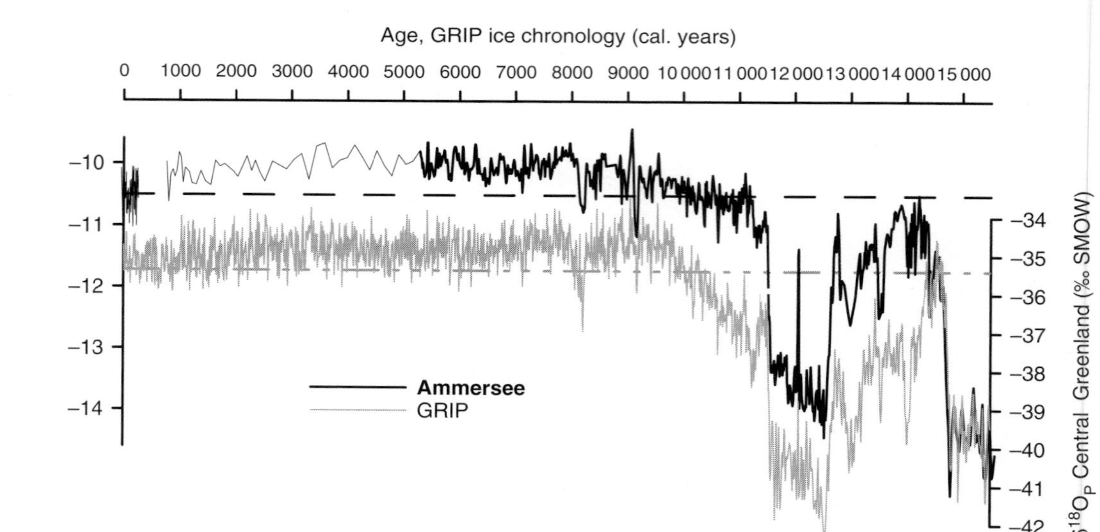

FIGURE 2.8. A comparison of the record from Ammersee, in southern Germany, and the GRIP ice core showing the close correlation between the Younger Dryas cold event from 12.9 to 11.6 kya at the two sites (from von Grafenstein *et al.*, 1999).

The suddenness of the Bølling warming coincides with evidence of a rapid rise in sea level that started around 14.6 kya (see Section 2.10). This rise must have been the result of a significant collapse of one of the major ice sheets. A warming resulting from an influx of freshwater is not consistent with an event in the North Atlantic, as this would have resulted in a cooling similar to the Heinrich event 1. Current thinking is that this water came from the break up of part of the Antarctic ice sheet (Clark *et al.*, 1996; Weaver *et al.*, 2003). This surge of melting ice led to a dramatic cooling of the Southern Ocean and had the counterintuitive effect of switching on the thermohaline circulation in the North Atlantic (see Section 2.11), and a rapid warming of the northern hemisphere.

Together with Heinrich event 1, the Bølling was part of the paroxysmal changes that occurred between the LGM and the Holocene, which were associated with the collapsing of the great ice sheets in both hemispheres. This interstadial was, however, relatively short-lived. Within a few hundred years it suffered a sharp interruption (known as the *Older Dryas*) before returning to a somewhat warmer level (this interstadial, the *Allerød*, is named after the place in Denmark where the first evidence of its existence was identified). Then around 12.9 kya the climate dropped back into near glacial conditions for over a thousand years, known as the *Younger Dryas*. It has been argued that the scale of this event justifies its being defined as an additional Heinrich event. There is, however, little evidence of its having been associated with a surge of icebergs (see Fig. 2.7) and it is more likely that the release of a large amount of glacial meltwater from North America (see Section 2.10) was the cause of this substantial cold event.

The Younger Dryas can be regarded as the last great shudder of the ice age and affected the entire northern hemisphere. Measurements from sediments off the coast of Venezuela (Haug *et al.*, 2001) to lakes in Ontario (Yu & Eicher, 1998) and southern Germany (von Grafenstein *et al.*, 1999; Fig. 2.8) all present the same story: a sudden dramatic drop in temperature that remained in place for nearly a thousand years, followed by an equally sudden warming. In Europe during this frigid spell summer temperatures were 5 to 8 °C below current values, and in midwinter they were 10 to 12 °C lower. They then returned rapidly to temperatures comparable to those of the Allerød at the beginning of the Holocene. There was, however, a temporary interruption. Known as the *Preboreal Oscillation*, this short-lived cooling, lasting some 200 years, occurred around 11.25 kya. Originally discovered in northwestern Europe, this brief cold episode is clearly seen in the Greenland ice cores and in some North Atlantic marine records, and somewhat less extensively in eastern North America.

When viewed on a global basis there are a number of odd features about the end of the last ice age. Around Antarctica there is little evidence of the dramatic ups and downs that marked the emergence of the northern hemisphere from the icy grip of the glacial. More striking is that the warming started two to three thousand years earlier in the southern hemisphere (Blunier & Brook, 2001). This delay may be related to the presence of the huge thermal inertia of the ice sheets on the northern continents, which slowed things down at higher latitudes for several millennia. This is hardly surprising if you think about how long it would take for the ice to melt. Even allowing for the dramatic changes in atmospheric circulation that may have occurred in the northern hemisphere, together with the increased summer insolation at high latitudes, a simple calculation of the time needed to supply the amount of energy needed to melt the ice sheets leads to a figure of several thousand years.

Until around 10 kya, then, the global climate was continually adjusting to a series of huge shifts. Thereafter, the climate settled into what looks like an extraordinarily quiet phase, when viewed against the upheavals that went before. There was still the matter of the final disappearance of the northern ice sheets. Over northern Europe, the Fennoscandian ice sheet largely disappeared by around 8.5 kya after creating a series of meltwater lakes in the region of what is now the Baltic Sea (see Section 2.10). The more massive Laurentide ice sheet did not completely disappear until around 6 kya. Furthermore, its death throes appear to have produced one last, short-lived, sharp cooling of about 5 to 6 °C over Greenland and more like 1.5 to 3 °C around the North Atlantic region between about 8.2 and 8 kya. This temporary setback has been linked to the sudden release of the floodwaters of two meltwater lakes that had collected over central Canada behind the melting Laurentide ice sheet. This surge of freshwater out through the Hudson Strait and into the Labrador Sea appears to have disrupted the circulation of the North Atlantic and precipitated one last hemispheric cold event (Barber *et al.*, 1999; see Section 2.10).

2.7 THE HOLOCENE

The transition from the Younger Dryas to the Holocene marks the break from the turbulent climate that characterised the ice age to something far more benign. The timing of this change is a matter of definition. The temperature rose to something comparable to modern values by around 10 kya. The subsequent 200-year cooling between around 8.2 and 8 kya described in the previous section is seen as a relic of the last ice age. Apart from this hiccup, the ice-core temperature record suggests the Holocene was a period of remarkable climatic stability. When compared with the previous 90 kyr this is a reasonable assessment, but we must not throw out the baby with the bathwater. Within the impression of relative calm and stability there lurk some important fluctuations. These serve to put the awfulness of the climate of the ice age into context because, in spite of their limited scale, they had a profound impact on human history.

Starting with the broad features of the Holocene, the global climate reached an optimum around 6 kya. Just how benign the climate became can be judged by the expansion of forests to high latitudes of the northern hemisphere. They reached their limits across Eurasia around 7 kya, some 200 to 300 km farther north than their current extent. In northern Canada this expansion was delayed a further two to three thousand years by the slower collapse of the Laurentide ice sheet, which did not completely disappear until around 6 kya, when the post-glacial warming reached its peak. On the evidence of tree cover the average summer temperature in mid-latitudes of the northern hemisphere was 2 to 3 °C warmer than it is today. Not only had trees spread farther north than now but they also extended higher into upland areas: in Britain trees grew at levels 200 to 300 m above the current timberline. This phase of the climate is known as the 'Atlantic' period, because it is assumed to have featured a strong westerly circulation which brought warm wet conditions to high latitudes of the northern hemisphere.

At lower latitudes the warmer conditions showed up most noticeably in a stronger summer monsoon circulation. This brought heavier rainfall to many parts of the subtropics. It affected not only the Indian subcontinent but also the Middle East and much of the Sahara. Ice cores from glaciers high in the Peruvian Andes confirm that the period from 8 to 5 kya marked the climatic optimum in the tropics (Thompson *et al.*, 1998). All this appears to provide a consistent global picture of stability in tune with the evidence from the Greenland ice-core temperature records, and many of the proxy records in mid-latitudes. In the tropics, however, there is one particular fly in the ointment: the desiccation of the Sahara. Here, the moister conditions that had lasted since around 14.5 kya suddenly came to an end around 5.5 kya (deMenocal *et al.*, 2000) . This dramatic shift has to do with the location of the region of rising warm moist air and cloudiness that encircles the globe in equatorial regions. Known as the Intertropical Convergence Zone (ITCZ), it governs many features of the global climate and appears capable of responding suddenly to longer-term weather cycles (see Section 2.11).

Measures of dust and sea salt in the Greenland ice cores during the Holocene show rather more variation and suggest that the strength of the mid-latitude westerlies varied considerably on millennial time-scales. So it may be that these aspects of global weather patterns explain why in some parts of the world there were more dramatic shifts than seen in the Greenland temperature records. This takes us into the more subtle areas of latitudinal shifts in global weather patterns and the non-linear response of the climate to long-term changes in the mid-Holocene. At the same time as the Sahara began to dry out, southern Africa became warmer and moister and the Antarctic pack ice shifted to lower latitudes (Hodell *et al.*, 2001). A sediment core from the South Atlantic, which is now close to the limit of pack ice cover, showed that between 10 and 5.5 kya there was virtually no ice-rafted detritus in the core. Thereafter, pack ice returned to these latitudes.

Another interesting change was that the quasi-periodic fluctuations in the tropical Pacific associated with the weather pattern known as *El Niño* appear to have been largely absent during the early Holocene (Rodbell *et al.*, 1999; Sandweiss, Maasch & Anderson, 2001). These warm events in the equatorial Pacific are part of the meteorological phenomenon El Niño/Southern Oscillation (ENSO) – a quasi-cyclic oscillation resulting from ocean–atmosphere coupling across the Pacific. It switches between periods of above-normal sea surface temperatures (SSTs) in the eastern equatorial Pacific and below-normal SSTs in the west (El Niño conditions), and the reverse (La Niña conditions).

A record of SSTs obtained from magnesium/calcium ratios in foraminifera from seafloor sediments near the Galápagos Island since the LGM provides insights into how ENSO fluctuations may have influenced the global climate (Kavoutas *et al.*, 2002). The observed LGM cooling of just 1.2 °C implies a relaxation of tropical temperature gradients, a southward shift of the ITCZ, and a persistent El Niño pattern in the tropical Pacific. Temperatures then fluctuated over a range of about 1 °C from the end of the LGM to the early Holocene in line with larger changes in the northern hemisphere. During the mid-Holocene there was a cooling of nearly 1 °C, which suggests a La Niña pattern with enhanced SST gradients and strengthened trade winds.

An ice core from high in the Bolivian Andes (Thompson *et al.*, 1998), which also provides data extending back into the last ice age, shows that El Niño events returned around 5.5 kya. There was a clear shift to more variable conditions that can be attributed to the return of the El Niño around this time. This change is supported by archaeological evidence of shells of molluscs along the coast of Peru, and by detailed cores taken from the bottom of Laguna Pallcacocha in Southern Ecuador (Rodbell, *et al.*, 1999). The latter appear to show a link with the 1500-year periodicity in the ocean sediment records from the North Atlantic that may be linked to DO events (see Section 2.5).

What is particularly interesting is that the sudden changes in the Sahara appear to reflect a response to the lengthy cycles associated with the variations in the Earth's orbit that have driven the 100-kyr cycle in the ice ages over the last million years or so (Section 2.11). What appears to have happened is that the ITCZ moved southwards. Any shifts in its position or intensity can have a crucial impact on how wet or dry different areas are. So, if the monsoon rains that brought relatively heavy rainfall to much of the Sahara region during the early Holocene ceased to extend so far north, then the Sahara would rapidly become desert. This is what appears to have happened in the central and eastern Sahara, and in Arabia, where extreme aridity took over within a century of so. The fact that this sudden shift coincided with changes in the tropical Pacific emphasises the global nature of connections in the tropical climate. Changes over Africa and Arabia, as well as in El Niño conditions, are part and parcel of movement in the position of the ITCZ.

Beyond the tropics the mid-latitude storms belts appear to have moved a few degrees of latitude towards the Equator. In these regions the changes were less profound, although, as noted earlier, the treeline across the northern continents receded southwards 200 to 300 km over subsequent millennia. Observations of changes in the amount of sodium ions (a measure of sea salt) and potassium ions (a measure of continental dust) in the Greenland ice cores (Mayewski *et al.*, 1997) suggest that both the Icelandic low and Siberian high intensified between 5.8 and 5.3 kya. This indicates that the circulation in mid-latitudes of the northern hemisphere strengthened at this time.

These dramatic changes in the mid-Holocene show how closely we have to look at regional records if we are to get an accurate picture of shorter-term changes in the climate. Evidence of these changes has recently been collated as part of a concerted international assessment, involving universities from the United States, United Kingdom, Sweden and South Africa. Known as the CASTINE project (for Climatic Assessment of Transient Instabilities in the Natural Environment),[4] it has identified at least four global periods of rapid climate change during the Holocene. These include the two periods we have already touched on, around 9 to 8 kya, and 6 to 5 kya, plus periods between 3.5 and 2.5 kya and since 0.6 kya. These more turbulent periods show up in the form of stronger mid-latitude circulation in the northern hemisphere, expansion of mountain glaciers around the world and greater sea-ice formation in the northern North Atlantic, as seen in the amount of ice-rafted debris found in ocean sediments. In addition, there were two more regional periods of change around 4.2 to 3.8 kya and 1.2 to 1.0 kya.

This book concentrates on the periods before 2.5 kya. Although the event 1.2 to 1.0 kya is thought to have included the drought in Central America that precipitated the collapse of the Mayan civilization (Haug *et al.*, 2003), for current purposes it is too recent to be presented as part of prehistory. This criterion most certainly applies to the period since the Middle Ages, which is usually referred to as the Little Ice Age, and cannot be squeezed into prehistory. As for the more detailed features of the changes that occurred in different parts of the world during the earlier climatically turbulent periods, these will be discussed in terms of their potential implications for interpreting prehistoric events as and when they are dealt with in later chapters.

2.8 CHANGES IN CLIMATE VARIABILITY

So far we have concentrated on the major changes in the climate during the last 100 kyr. This is only part of the story. As noted in Section 2.1, changes in the shorter-term variability of the climate are just as important. The GISP2 ice-core record since 60 kya has been sampled for 20-year intervals over its entire length (Fig. 2.9(a)). What

[4] Correspondence on the CASTINE project should be addressed to paul.mayewski@maine.edu.

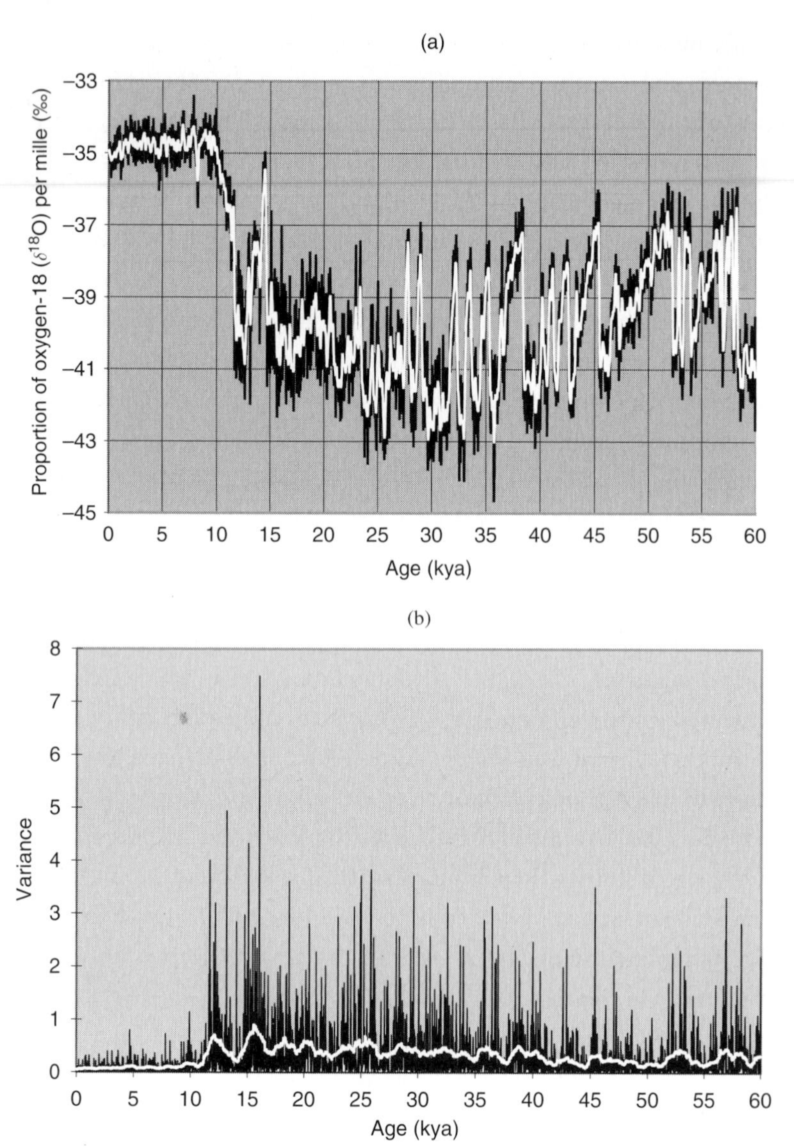

FIGURE 2.9. An estimation of the change in the variance of the climate over the last 60 thousand years. In (a) the GISP2 ice-core data for every 20 years (black curve) is overlain by the 21-term binomial running mean (white curve). In (b) the square of the difference between the two curves in (a) is presented, together with its long-term variation (white curve). This presentation provides a measure of the variance of the climate in the vicinity of Greenland and the North Atlantic. (Data archived at the World Data Center for Paleoclimatology, Boulder, Colorado, USA.)

we see, superimposed on the ups and downs during the last ice age and the subsequent warming, is the fuzz of shorter-term climatic variability. This fuzziness is a measure of the changes from decade to decade (i.e. well within a human lifetime). How this variability has changed over the last 60 kyr can be extracted from the ice-core records. This is done by computing the squares of the difference between the individual measurements (the black line in Fig. 2.9(a)) in the record and the smoothed version of the data that removes periodicities shorter than about 200 years (the white line in Fig. 2.9(a)). Known as the *variance*, this measure provides an indication of the disruptive potential of climatic fluctuations at any given time.

The use of the square of the deviation from normal to provide a measure of the impact of extreme weather events may seem like a statistical sleight of hand. In fact, it can be supported by a rather simple example. In considering the economic impact of tropical storms, it has been shown that the damage they cause when they strike land is roughly proportional to the square of the wind speed. So in keeping a tally of trends in damaging storms a measure of the square of their strength is used to give a more accurate indication of their potential to cause economic damage. More generally, there is widespread evidence that bigger deviations in the weather have a disproportionately greater impact on many human activities.

Using the square of the deviation of the short-term fluctuations from the longer-term trend in the GISP2 record shows how much greater the variability of the climate was before about 10 kya, and the relative stability since then (Fig. 2.9(b)). The striking feature of this calculation is the huge decline in the variance at the end of the last ice age. Whereas the entire period from 60 kya until around 12 kya is marked by extreme variability, thereafter the climate settled down into a much more quiescent pattern. Broadly speaking the variance dropped by a factor of five to ten. This change can be described as the statistical equivalent of the world's emerging from the climatic 'long grass' into more hospitable conditions. This approximately order of magnitude drop in variability is a vital factor in the analysis of the ability of humans to develop more settled societies, a theme we will return to throughout this book.

Other measures of rapid fluctuations come up with the same broad conclusions. The rate of accumulation of the ice cores, which can be measured for individual years since around 15 kya, shows the annual rate of snowfall as varying by a factor of three in a few years until around 10 kya (Alley *et al.*, 1993), whereas thereafter it was much less variable. In addition, the observations of wind-blown dust and sea salt, which provide measures of both the strength of circulation in mid-latitudes and, in the case of dust, the aridity of the continents in these latitudes, present if anything an even more tempestuous picture of extreme variability until the start of the Holocene and the quiescence of the last 10 kyr.

In terms of the conditions that humans had to survive, however, the transition from chaos to relative stability was a matter of the rapidity and size of local fluctuations. As with longer-term changes, we need to check whether extrapolation from one part of the world to another is acceptable. In Section 2.4 it was argued that, on the basis of proxy records from around the world, it was reasonable to use data from Greenland to draw conclusions about changes in the climate worldwide. The case for extending the analysis to shorter-term variability is also based on the fact that, in terms of mid-latitude circulation patterns, the observations obtained from the Greenland ice cores are closely linked to weather conditions around the northern hemisphere.

When it comes to the tropics the position is less clear. There is, however, one limited tropical source of information about shorter-term climatic variability at low latitudes during the last ice age. This comes from studies of the isotope ratios of ancient corals found on the Huon Peninsula, in Papua New Guinea (Tudhope *et al.*, 2001), that provide insights into both the sea surface temperature and the strength of ENSO, which is a dominant influence on the distribution of rainfall throughout the tropics. Measurements on corals laid down around 40 and around 85 kya show that SST at those times was 2 to 3 °C cooler and that ENSO interannual variance was less than in modern corals. This pair of snapshots imply that the climatic variance in the tropics during much of the last ice age was less than in modern times. The somewhat contradictory conclusion is that, while the climate was much more variable in the higher latitudes of the northern hemisphere, it may have been less volatile in the tropics.

This conclusion is supported in part by a basic feature of the current global climate. Computer models support the general hypothesis that if the temperature difference between the tropics and polar regions increases, the circulation will become stronger and stormier. Although the tropics cooled by a few degrees during the ice age, the drop in temperature was much greater at high latitudes. This would have wound the circulation up in mid-latitudes, making it more variable. So it is probably reasonable to assume that the dramatic shift in climatic variability at the beginning of the Holocene was a hemispheric phenomenon outside the tropics. This means that until around 10 kya, for all but the tropics, the climate was much more variable than it has been since then.

The importance of the distinction between climatic variability and longer-term changes cannot be underestimated. It is true that the transition from the end of the Younger Dryas into the Holocene brought warmer and wetter conditions to much of the northern hemisphere. The scale of these changes was not, however, that great at lower latitudes. Furthermore, the regional changes were, as we will see in later chapters, complicated. So, while the relatively rapid changes between the LGM and the beginning of the Holocene had a dramatic impact on humans, especially in Eurasia and the Middle East, it was the fundamental drop in variability that was to tip the balance in favour of agriculture. In short, using the factor of five to ten increase in variance as a yardstick, agriculture, as we know it, would be impossible in a world where conditions varied so much (Richerson, Boyd & Bettinger, 2001).

2.9 JUST HOW CHAOTIC IS THE CLIMATE?

In drawing a statistical distinction between conditions during the last ice age and in the Holocene, a measure of variance is used to assert that the climate was more chaotic during the former period. This approach appears to be reasonable in terms of considering the implications of shifting climatic variability on humankind. There is, however, a more fundamental issue of whether it is correct to talk in terms of the climate being 'more chaotic' during the ice age and less so since. This is not simply a matter of changes in decadal variance, but whether the general behaviour of the climate is in any way predictable. This extends the analysis to the sudden changes in the ice-age climate that were described in Section 2.5. Such dramatic shifts suggest much more unpredictable behaviour, with implications for both our analysis of the past, and our predictions of future climate change.

Examining whether the climate is a chaotic system is a matter of looking at a wider range of timescales. The atmosphere is a turbulent fluid and weather systems behave in a chaotic manner. Experience tells us that, in spite of the massive power of modern computers, numerical weather forecasts lose much of their skill beyond about a week. But, although the atmosphere is chaotic on a day-to-day basis, the same need not apply to longer-term averaged conditions. Within the broad bounds of the annual cycle the climate in any particular part of the world generally sticks within prescribed limits: the temperature hardly ever rises above –20 °C at the South Pole, or falls below 20 °C in Singapore. The oceans interact with the atmosphere in a way that enables us to use knowledge of their slowly varying characteristics to make useful predictions of seasonal weather. In addition, the variations associated with the ice ages on timescales of tens of thousands of years, which were described in Section 2.3, can be largely explained in terms of changes in the Earth's orbital parameters (see Section 2.11). These results imply that some features of the climate are largely predictable.

It is when we turn to the variations between decades and millennia that the real differences can be seen between the glacial and interglacial climates of the Earth. It is not just the substantial difference in the interannual and interdecadal climatic variability, but also the sudden shifts that, within a few years, switched the global climate into substantially colder or warmer conditions that then lasted for periods from 1500 to 5000 years. Clearly, the behaviour of the glacial climate was far less predictable than the climate of the Holocene. So, it is not an exaggeration to say that the climate during the ice age was far more chaotic than now. Moreover, the suddenness of the change around 10 kya had profound implications for humankind. What is more, any return to more chaotic conditions could have equally significant implications for our future.

2.10 CHANGES IN SEA LEVEL

In 1931 a trawler working in the southern North Sea dredged up a lump of peat containing an exquisitely crafted spearhead made from a deer's antler (Mithen, 2003, pp. 150–1). Dated as being nearly 14 kyr old, this artefact was dramatic evidence of how early humans exploited the broad expanses of land that had been exposed during the last ice age, and were only reclaimed by the sea some 7 kya. When this spearhead was buried, dense oak forests had yet to spread into the region, known to archaeologists as 'Doggerland', where now the sea is over 30 m deep. This famous find emphasises that the rise in sea level between about 15 and 5 kya covered up large areas of habitable land that had been exploited by humans and made movement around the continental margins easier.

There is an additional reason for considering the rise in sea level. Even today changes in sea level raise powerful emotional reactions. On the southeast coast of England the chalk cliffs are seen as part of this country's heritage. In 2000 a large chunk of the most spectacular part of these cliffs, at Beachy Head, fell into the sea. This event was seen as clear evidence of rising sea level and was greeted with a great sense of powerlessness and foreboding about the consequences of global warming. Although there is reason to worry about rising sea level, the fact that the changes occurring around the coasts of Britain

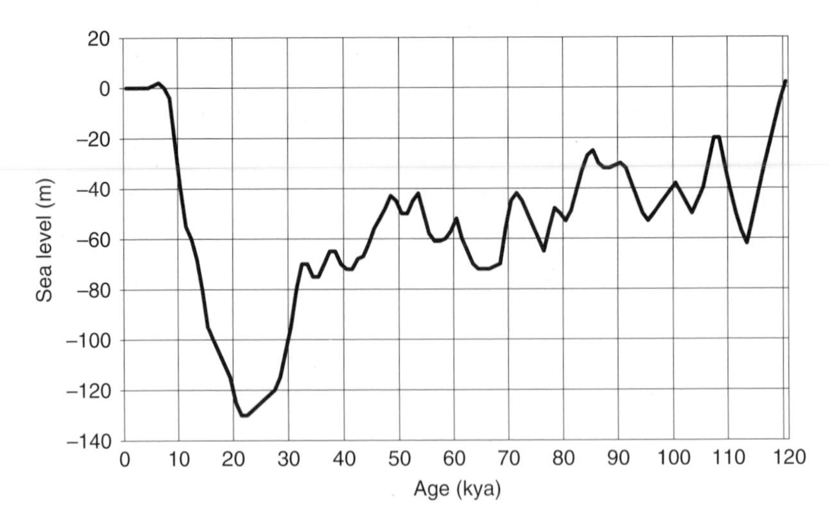

FIGURE 2.10. Changes in sea level during the last 100 kyr, based on Mix, Bard and Schneider (2001), Lambeck *et al.* (2003), and Lambeck and Chappell (2001).

are principally part of the adjustment that has been going on since the end of the last ice age must influence our thinking. Some eight thousand years ago it was still possible to walk to France on dry land.

Because the scale of the changes in sea level during the last ice age constitute such a massive change in the landscape we need to include the scale and timing of these changes in our thinking on human development. At the end of the last interglacial around 117 kya the sea level was as high as or higher than in current times. Then the level fell sharply by some 50 m by around 112 kya (see Fig. 2.10; Lambeck, Esat & Potter, 2003). Thereafter it fluctuated between about 30 and 70 m below current levels during subsequent interstadials and stadials until the end of OIS3 around 30 kya. Although these fluctuations may not appear that substantial when compared with low levels reached during the final stages of the ice age, they may have had a significant influence on how humans migrated across the globe. In particular, the drop after 85 kya, and the low level between about 67 and 61 kya, are important in this context.

The scale of the sea-level fluctuations tended to decline during OIS3, as the ice sheets built up. The level then plunged at the beginning of OIS2, and during the LGM was around 120 m below current levels. These are massive changes, and completely altered the geography of the continental margins of many parts of the world. From northwest Europe to the eastern Mediterranean, Indo-China, between Siberia and Alaska (a region usually known as 'Beringia') and the eastern seaboard of North America, huge lowland areas were exposed for tens of thousands of years. Here modern humans would have been able to exploit the rich resources of the coastal plains.

One of the great unknowns of any analysis of human social development during the warm conditions after the Younger Dryas is just how much evidence has been swept away by the rise in sea levels. If, as seems likely, some of the most stable and best-fed communities would have quickly developed close to seashores, any evidence of their existence will have long gone. The simple fact of the matter is that this rise in sea levels drowned or washed away evidence of nearly all earlier coastal adaptation everywhere around the world.

When it comes to making a detailed analysis of sea-level rise in any given locality, our efforts are often complicated by other factors. In particular, *isostatic uplift* and *tectonic movements* may make life difficult. Isostatic uplift occurs where the melting of ice sheets since the end of the last ice age has removed a load from the Earth's crust, which then rises up out of the underlying mantle. In places like Scotland, Scandinavia, Alaska and Newfoundland this has had, and is still having, a significant impact on the coastline. For instance, the crust in the region around Stockholm has risen by 80 m in the last 5 kyr and is still rising at a rate of nearly a metre a century. But, although inexorable and substantial, these changes make little difference to how we interpret human prehistory. The movement of tectonic plates that make up the Earth's crust (often known as *continental drift*) also leads to changes in the apparent level of the sea. These movements have greatest impact in active earthquake zones such as the Mediterranean, Japan and the west coast of North America.

The waxing and waning of the ice sheets in the northern hemisphere had a variety of additional consequences for the landscape and also for the global climate (see next section). Two landscape effects are of particular interest. First, the melting ice, especially after the LGM, produced large bodies of water that were sometimes cut off from the oceans for hundreds of years. Second, as the ice melted, the crustal uplift altered the regional geography and sometimes contributed to the formation of long-lasting bodies of water.

The combination of these effects is most difficult to interpret over northern Europe. The fossil evidence shows that as the Fennoscandian ice sheet started to melt around 14.5 kya, a freshwater lake, known as the 'Baltic ice lake', initially flooded the region to the south of the ice. The value expanded until around 12.9 kya when the ice receded sufficiently for some 7000 km^3 of freshwater to flood out across the depressed central southern Sweden and then to flow out into the North Sea. A glacial readvance during the Younger Dryas blocked this drainage route and it was not until the climate warmed around 11.5 kya that the final drainage of the Baltic ice lake occurred. Then the rising sea level meant that water flowed in the reverse direction and a saltwater body formed (named the Yoldia Sea after a characteristic saltwater fossil). The new sea lasted for about 500 years, before the recoil of the crust in southern Sweden and northern Germany cut off the supply of saltwater and the lake to the south of the ice sheet returned to being freshwater. Named Ancylus Lake, after a fossil freshwater shell, this body of water lasted until around 8.5 kya, by which time the ice sheet had long gone. Then the rising sea level and crustal movements allowed a link to form between the North Sea and the lake, and saltwater flowed into it. This brackish body of water, named the Littorina Sea, after yet another fossil shell, eventually transformed into what is now the Baltic Sea. The shape of this sea has undergone major changes over the last 7 kyr or so, as the crust has continued to rise.

Across North America the story of ice-dammed lakes and catastrophic outbursts occurred on an even larger scale. Lake Agassiz was the largest of the proglacial lakes that formed in central North America during the retreat of the Laurentide ice sheet. It is reckoned that Lake Agassiz assumed a variety of forms during its 4.5-kyr history (around 13.7 to 8.2 kya). It underwent major changes in volume during several catastrophic outbursts by the opening of new (lower) lake outlets as the southern margin of the Laurentide ice sheet retreated. The changes in area and volume were considerable (Leverington, Mann & Teller, 2002). Its early stages had areas and volumes of up to about 170 000 km^2 and 13 000 km^3, respectively (about the volume of modern Lake Superior). The largest of the middle stages of the lake had areas and volumes of about 250 000 km^2 and 23 000 km^3 (about the total volume of the modern Great Lakes). The latest stages of Lake Agassiz were much larger than all preceding stages, as a result of the lake's merger with glacial Lake Ojibway. Just before the final catastrophic release of its waters into the Hudson Bay, the area and volume of the combined Lake Agassiz–Ojibway had grown to 841 000 km^2 and 163 000 km^3. At this time, the lake's volume was about seven times the total volume of the modern Great Lakes.

Among the largest catastrophic meltwater pulses from Lake Agassiz into the North Atlantic were those at 12.9 kya (9500 km^3), 11.3 kya (9300 km^3), and 8.2 kya (163 000 km^3). These outbursts coincide with the start of the Younger Dryas, the Preboreal Oscillation, and the 8.2 kya event, suggesting that outbursts from Lake Agassiz may have repeatedly influenced hemispheric climate by affecting the circulation of the North Atlantic. This, in turn, altered the temperature of the surface of much of the northern North Atlantic, and with it the climate of much of the northern hemisphere (Barber *et al.*, 1999).

In the northwest United States where the Cordilleran ice sheet extended into northern Idaho and Montana, it dammed the Columbia River valley. Behind this ice a glacial lake, known as Lake Missoula, existed between around 15.3 and 12.7 kya. When the glacial dam collapsed, as it did many times during its lifetime, it released some 2000 km^3 of water. This rampaged westward at 100 kph in a torrent that completely devasted the landscape of the narrow river valleys along which it flowed.

Farther south, what is now the Great Salt Lake was much larger and formed a lake called Lake Bonneville that, at its peak, had a surface area of 51 000 km^2. The lake built up between 30 and 15 kya, reaching a maximum around the end of the LGM. The waters for Lake Bonneville were not glacial melt, but a consequence of regional weather patterns that brought frequent depressions and abundant rainfall to the Great Basin. Around 14.5 kya the lake level abruptly dropped by 100 m, producing the Bonneville Flood, as a result of catastrophic failure of the natural dam that had retained the maximum lake level. This released about 5000 km^3 of water, in a matter of days, into Colorado River, some three orders of magnitude greater than the most extreme modern floods. During the Holocene more arid climatic conditions have systematically lowered the lake levels, leading to today's residual in the Great Salt Lake.

For anyone living in the vicinity of flood events of this magnitude the consequences would have been catastrophic. Farther afield, the biggest of these outbursts may have played an important role in human settlement and migration along shallow continental shelves and marine basins such as the Persian Gulf, perhaps pushing already stressed coastal systems over a threshold. For example, the 163 000 km^3 meltwater pulse from Lake Agassiz at 8.2 kya would have caused an sudden incursion of the sea by 1 km inland across the floor of a gentle continental shelf with a slope of 1 in 2000, or a substantial advance of more than 10 km in only about a year across almost level land. While the rise in postglacial sea level may not have been sufficient to generate a folk memory of the 'Flood Myth', the largest outburst from Lake Agassiz at 8.2 kya may have made a more indelible impression (see Section 5.15).

A similar situation developed in Siberia during the LGM. Here single giant freshwater lake formed. It stretched some 1500 km from north to south, and a similar distance east to west at its widest points. At its maximum extent it would have had a surface area at least twice that of the Caspian Sea. It was formed by the ice sheets to the north damming off the Yenisei and Ob rivers. This huge lake appears to have reached its greatest extent by around 24 kya, and to have existed in some form until around 14 to 16 kya. There is, however, little evidence of its draining away in the same catastrophic manner as the glacially dammed lakes in North America.

Much less is known about the fluctuations in sea level arising from the reduction in size of the Antarctic ice sheet. As noted in Section 2.6, the first major sudden rise in sea level after the LGM was the one that occurred around 14.6 kya. This rise of some 20 m, in no more than a few centuries, is now attributed to events in Antarctica (see Section 2.11).

2.11 CAUSES OF CLIMATE CHANGE

There is not space here to go into all the proposed explanations of the changes that have occurred in the last 100 kyr: there are more than enough textbooks on the subject (see Bibliography). All that is needed here is a brief summary of how virtually all the observed forms of change set out in Section 2.1 can be explained in terms of known physical processes. Starting with the broad cycle of the ice ages, which occur every 100 kyr or so, and have lesser cycles of around 20 and 40 kyr, received wisdom is that they are explained remarkably well in terms of the Earth's orbital parameters (Imbrie *et al.*, 1992, 1993). These modify the amount of sunlight falling at different latitudes at different times of the year. It is found that the key to explaining how variations in the orbital parameters can trigger ice ages is the amount of solar radiation received at high latitudes during the summer. This is critical to the growth and decay of ice sheets. At 65° N this quantity has varied by more than 9% during the last 800 kyr. Computer modelling studies show that, when fluctuations of this order are combined with realistic assumptions about the time taken for ice sheets to build up, it is possible to reproduce the long-term behaviour of the ice ages with surprising accuracy. These calculations can reproduce the broad form of the climate changes measured in ocean sediments and the EPICA ice core (see Fig. 2.5), with its saw-tooth pattern of short interglacials every 100 kyr or so, followed by a slow descent into full glacial conditions and then a sudden warming into the next interglacial. Superimposed on this pattern are cycles of around 20 kyr in duration.

The fact that the orbital theory of ice ages is measured in tens of millennia means that these changes form the background to what follows. For the most part this is all we need to say, but for one particular feature. This is the nature of the Holocene variability in the Sahara. In the current interglacial the peak in summer insolation at 65° N occurred around 10 kya. In particular, rainfall over the Sahara during the past 20 kyr has exhibited a remarkable non-linear sensitivity to these gradual changes in insolation (see Section 2.7). What is particularly interesting is that the sudden changes in the Sahara appear to reflect a response to the lengthy cycles associated with the variations in the Earth's orbit.

Understanding changes in the Sahara requires us to look at tropical circulation around the end of the ice age. During the LGM much of northern Africa was, as now, arid desert. If anything this desert was even more extensive as the tropical rain forests were greatly reduced in extent. The reason for this shift has to do with the strength of the Intertropical Convergence Zone (see Section 2.7). This rising air produces heavy rainfall and the now dry air then spreads north and south before descending at around 20° N and 20° S. The very dry descending air then flows back towards the Equator. The entire circulation pattern is known as the Hadley cell. During the LGM the tropical oceans were cooler and the Hadley cell was less vigorous.

The more arid conditions during the LGM show up as a high level of wind-blown mineral dust in the sediment cores from the northern tropical Atlantic. Around 14.5 kya the level of dust dropped suddenly, suggesting the onset of much moister conditions (Fig. 2.11; deMenocal *et al.*, 2000). The amount of dust blown westwards into the Atlantic remained low apart from an increase during the Younger

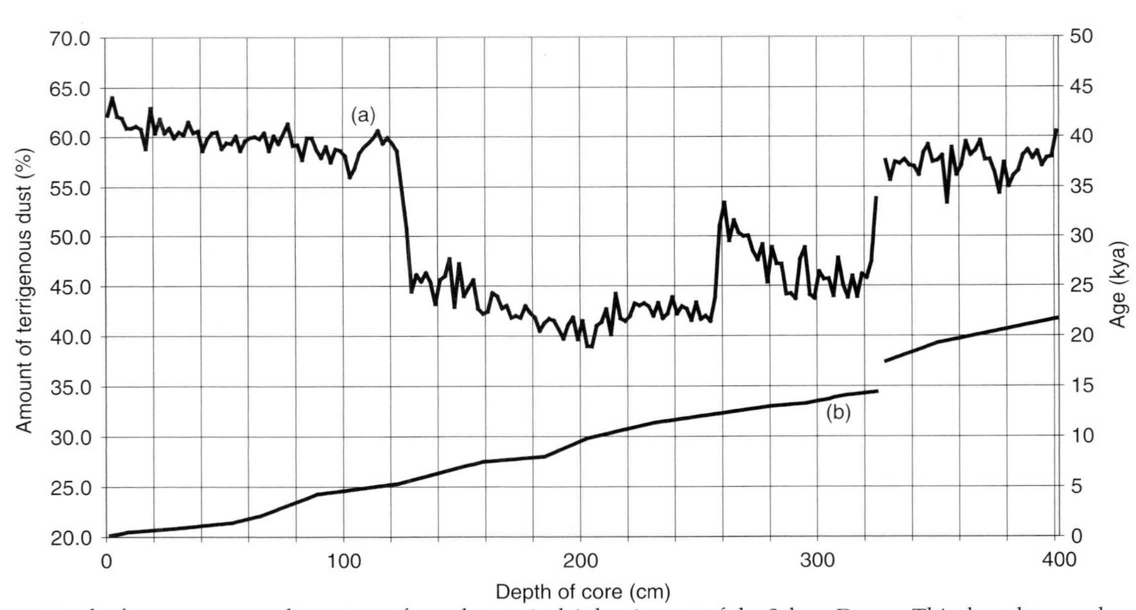

FIGURE 2.11. Results from an ocean sediment core from the tropical Atlantic west of the Sahara Desert. This data shows a sharp decline in the amount of mineral dust being transported from the Sahara between around 15 kya and 5 kya (from deMenocal *et al.*, 2000, using data available on Peter deMenocal's website).

Dryas and then a return to moister conditions. These moister conditions lasted until around 5.5 kya when the sediment records show a sudden increase in the amount of dust being transported in the winds from the Sahara. Abruptly the climate shifted to a drier form and the desert began to take over.

What appears to have happened is that the cycles in solar insolation influenced the position of the ITCZ. Any shifts in its position or intensity can have a crucial impact on how wet or dry different areas of the tropics are. Around 14.5 kya the level of summer insolation rose to a critical level at 65° N: sufficient to trigger a sudden shift northwards of the ITCZ into the Sahara during the summer. This brought relatively heavy rainfall to much of the Sahara. The level of summer insolation remained sufficient to maintain this pattern well into the Holocene, except during the Younger Dryas when the disruption of North Atlantic circulation was sufficient to override it. Then around 5.5 kya the summer insolation at 65° N fell back below the critical level, the ITCZ moved south, and the Sahara rapidly became desert. This change was most rapid in the central and eastern Sahara, plus Arabia, where extreme aridity took over within a century on so. The fact that this sudden shift coincided with changes in the tropical Pacific may well have reinforced the scale of climatic change over Africa and Arabia as El Niño conditions tend to be associated with southward movement of the ITCZ.

Almost all the other features of the climate can be explained in terms of the natural variability of the global weather system. On timescales of millennia and shorter, various components of the climate system can interact to produce major fluctuations. In particular, the sudden changes identified as DO and Heinrich events in Section 2.5 appear to be linked to switches in the large-scale movement of water in the North Atlantic. These shifts are an example of the process driving the deep waters of the oceans, known as *thermohaline circulation* (Broecker, 1995). This results from changes in seawater density arising from variations in temperature and salinity. Where the water becomes denser than the deeper layers it can sink to great depths. The temperature depends on where the surface waters come from and how much heat the oceans either pick up or release to the atmosphere in both sensible heat and evaporative loss. The salinity of a given body of water depends on the balance between losses through evaporation and gains from either rainfall, or freshwater run-off from rivers and melting of the ice sheets of Antarctica and Greenland plus the pack ice of the polar oceans. In practice there are few regions where sinking waters have a major impact. *Deep waters*, defined as water that sinks to middle levels of the major oceans, are formed only around the northern fringes of the Atlantic (North Atlantic Deep Water). *Bottom waters*, which constitute a colder denser layer below the deep waters, are formed only in limited regions near the coast of Antarctica in the Weddell and Ross seas.

The changes that matter here are those associated with the North Atlantic. During and around the end of the last ice age the circulation here underwent sudden and substantial shifts. These led to the global climate being able to exist in distinctly different regimes even though the overall energy balance of the system had not changed appreciably. Heinrich events are most easily linked to changes of this type. During the last ice age the partial collapse of the ice sheet over North America, which caused a surge of icebergs to flood out into the North Atlantic, would have had a radical impact on ocean circulation. The combination of the huge influx of freshwater and the ice cover in winter would have effectively capped the northern part of the Atlantic and switched off any thermohaline circulation north of about 40° N.

The link between Heinrich events and changes in the northern hemisphere ice sheets appears to be well established. The climatic explanation of the first warming after the LGM, the Bølling (see Section 2.6) seems to have involved events in the southern hemisphere (Clark *et al.*, 1996; Weaver *et al.*, 2003). The influx of a large amount of meltwater into the Southern Ocean altered the formation of deep water around Antarctica and meant more tropical water moved northwards into the North Atlantic. This effect extended to high latitudes and led to a profound warming of not only the North Atlantic region but the northern hemisphere as a whole.

Although the most extreme events can be explained in terms of major adjustments in global ocean circulation, DO events rely on more complicated mechanisms. The most intriguing one is based on a stability analysis of the Atlantic Ocean during the ice age (Rahmstorf & Alley, 2002). In this approach there appear to be two circulation modes: one stable, and the other a weakly unstable one that lasted several centuries before spontaneously ending. In addition, there was a second unstable mode in which the formation of North Atlantic deep water is shut down completely, corresponding to a Heinrich event. This model suggests that the glacial Atlantic was an *excitable system*, in which a suitable perturbation could trigger a temporary transition of the state to an unstable circulation mode. In contrast a warm climate, like the present, is *bistable* and appears to be much less susceptible to disturbance. It is postulated that this process exhibits a threshold in respect of the background noisiness of the climate. Even today, this stochastic response to changes in the freshwater flux into the North Atlantic could have major climatic implications (see Section 8.5).

The importance of this proposal is that the available evidence suggests that the sensitivity of the climate to noise could have changed appreciably during and since the ice age. So it is possible that in the early warmer stages of the ice age, when the warm mode of the North Atlantic circulation was more stable, there were fewer and longer-lasting DO events. Conversely, at its nadir during the extreme cold around 20 kya, when the cold mode of circulation was more stable, there were also fewer DO events. In between, when the climate existed in a metastable state, random flickers between the two states might easily occur. If these flickers were in response to a weak 1500-year cycle, then they could take the form of a stochastic resonance between the two modes of North Atlantic circulation.[5]

[5] This phenomenon is described in Burroughs (2003), pp. 234–7.

There is some evidence of the existence of the 1500-year oscillation in ocean-sediment records for the Holocene (Bond *et al.*, 1997). These are, however, only very faint echoes of the DO events during the last ice age. Nevertheless, measurements of ice-rafted debris in ocean-sediment cores taken from various locations in the North Atlantic, notably between Greenland and Iceland, and to the west of Ireland, show evidence of changes in the circulation of the ocean during the Holocene. This debris originates from the east coast of Greenland and the Arctic archipelago Svalbard, and provides a measure of the strength of the Irminger current and the North Atlantic Drift (the extension of the Gulf Stream across the North Atlantic and into the Nordic Seas). The sediment records show periods of marked cooling at 2.8, 4.1, 5.9 and 8.2 kya. These changes in circulation correlate closely with measurements of solar variability, and suggest they could be the result of solar influences (Bond *et al.*, 2001). This has led to the proposal that the observed changes in the circulation of the North Atlantic are the result of stochastic resonance driven by solar variability (Rahmstorf, 2003).

Although these changes do not show up strongly in the Greenland ice-core records, apart from the 8.2 kya event, their implications for the climate in Europe and the Middle East must have been considerable. Significant shifts in sea ice extent would have had an impact on the North Atlantic Oscillation (NAO). In turn, this would have altered rainfall patterns across Europe (see Section 5.1). The cooler episodes would have reduced the amount of rainfall, and it is interesting that these periods do coincide with periods of drought and social breakdown in Middle East (see Chapter 6).

These observations raise one further important question about the stability of the Holocene: in what circumstances does the climate of the North Atlantic become metastable again? There is lively debate in the climatological community as to whether current global warming resulting from human activities could push the climate into a more erratic mode. Some computer models of the global climate suggest that this is a distinct possibility. It is a frightening thought.

The evidence of the last ice age shows that a less stable climate would pose daunting challenges for the human race. So both understanding the nature of current climate change and appreciating fully how modern society emerged from the chaos of the last ice age assume even greater importance.

2.12 THE LUNATIC FRINGE

There is one last important point to consider in this brief review of the climatic background to human prehistory. This is the mass of what might best be called 'crackpot theories' about the nature of climate change at the end of the last ice age and the parallel emergence of modern humans. Involving everything from the intervention of aliens, through cosmic cataclysms, to explanations of the true nature of Atlantis, it could be argued that such theories do not warrant consideration in a book that is seeking to review current peer-reviewed scientific work on prehistory, climate change and anthropology. After all, as we will see, the conventional debate on these topics is not short of vigour. But to ignore completely the swirling cloud of alternatives that encircles this subject might give the impression that we are not prepared to face up to the challenge of these hypotheses. Better to touch briefly on them with the object of indicating why they are not needed: there is more than enough good stuff in the conventional science to keep us occupied. Furthermore, the scientific establishment has some enthralling explanations of many of the tantalising questions involved in the issue of climate change and prehistory.

The first thing to get straight is that the wide variety of sources of climatic data constitutes a remarkably consistent picture of past events. There is no need to invoke additional catastrophes to explain either the climate record or the possible utter destruction of early civilisations. There is no evidence whatsoever that there have been cataclysmic cometary impacts, or sudden shifts in the crust across the face of the Earth in the last 100 kya. So the possibility that major early unidentified civilisations were swept from the planet by an extraterrestrial impact about 10 kya has no support in the standard climate records. The Greenland and Antarctic ice sheets did not collapse in 'a matter of weeks' during some awful cosmic catastrophe. All the data is consistent with a steady, albeit climatically eventful, transition as the massive ice sheets took several thousand years to respond to the warming resulting from the changes in the Earth's orbit. In some places rising sea levels may have posed dramatic challenges for coastal communities, but as a general observation they are not the stuff of cataclysmic destruction of vast prehistoric and now vanished empires.

Even more absurd is the idea that Antarctica was located at lower latitudes and hence was ice-free at some relatively recent date (such as 15 kya), before suddenly moving to its current polar location. Although this might seem like a convenient source for the Atlantis legend, the evidence is unequivocal. Antarctica has been covered by ice for about 25 million years. There is no evidence of the type of massive upheaval that would have been necessary to enable such a major repositioning of the Earth's tectonic plates to have occurred. More important, the EPICA ice core (Fig. 2.5) provides an unbroken record for more than 730 kyr. This shows a remarkable parallelism with ocean-sediment observations around the world, and confirms that, for at least this period, conditions in Antarctica have been unremittingly icy. Those looking for a convenient place to postulate where a precursor civilisation to Egyptians could have developed will have to look elsewhere. For the rest of us, it is time to get down to the sobering process of considering the challenges for humans of living through the ice age and its subsequent collapse. This involves first summarising the climatic factors that have been reviewed in this chapter.

2.13 CONCLUSION: A CLIMATIC TEMPLATE

The essential features of the climate as far as modern humans are concerned are contained within the last 100 kyr. The observations reviewed in this chapter are encapsulated in Table 2.1 and can be defined as a 'climatic template'. In terms of climate change, during the last ice age, this involves some 20 DO warming events between 75 and 15 kya. The most important of these warming events were in all

TABLE 2.1 *Climatic template for the North Atlantic and Eurasia*

Age (kya)	DO events (kya)	Heinrich events (kya)	Interstadials and other events (kya)
0–10			Holocene climatic optimum (5–6 kya)
			8.2 kya cold event
10–20	1 (14.5)	H0 (12.9)	Younger Dryas (12.9–11.6 kya)
		H1 (16.5)	Bølling warm stage (14.5 kya)
20–30	2 (23.4)	H2 (23.5)	Denekamp interstadial (30–25 kya)
	3 (27.4)		
	4 (29.0)		
30–40	5 (32.3)	H3 (32.0)	Hengelo interstadial (38–36 kya)
	6 (33.6)	H4 (39.5)	
	7 (35.3)		
	8 (38.0)		
40–50	9 (40.1)	H5 (47)	Moershoofd interstadial (46–44 kya)
	10 (41.1)		
	11 (42.5)		
	12 (45.5)		
	13 (47.5)		
50–60	14 (52.0)		Glinde (51–48 kya)
	15 (54.0)		Oerel (58–54 kya)
	16 (57.0)		
	17 (58.0)		
60–70	18 (62.0)	H6 (67.0)	Prolonged stadial from around 67 to 61 kya across Eurasia
70–80	19 (70.5)		
	20 (74.0)		
80–90	21 (84.0)	H7? (87.0)	Odderade interstadial (84–75 kya)

probability numbers 17, 14, 12 and 8 (see Fig. 2.6), which appear to coincide with the Oerel, Glinde, Hengelo and Denekamp interstadials. Of possibly greater importance are the six Heinrich events between around 70 and 16 kya that marked the periods of most intense cold during the ice age, culminating in the LGM between 25 and 18 kya. Following the most recent Heinrich event there was a sequence of more rapid climate changes involving the Bølling, Older Dryas, Allerød and Younger Dryas, before the climate really warmed up to usher in the Holocene around 10 kya. Since then climate changes have been minuscule compared with what went before. Nevertheless, this quiescent period has seen some interludes of limited but rapid climate change, including the sudden cooling around 8.2 kya, and the major shifts in the tropics between 6 and 5 kya, plus periods between 3.5 and 2.5 kya and since 0.6 kya. In addition, there were two more regional periods of change around 4.2 to 3.8 kya and 1.2 to 1.0 kya.

Overarching these major changes, there has been a more fundamental shift in short-term climatic variability. Prior to the end of the Younger Dryas around 12 kya the climate was substantially more variable than it has been since. In terms of human activities this feature of the climatic template represents a far more important challenge, especially in respect to agriculture. So our exploration of human prehistory must focus on the timing of the major changes in the climate, both during the ice age and to a lesser extent during the halcyon millennia of the Holocene, as these may hold the key to some of the more puzzling changes in the development of social and economic structures around the world. At the same time we need to keep in the forefront of our minds the radical shift in climatic variability which is a feature of the onset of the Holocene, as it is central to so much of what has happened during the last 10 kyr, and never lose sight of the opportunities that arose as we emerged from the climatic 'long grass'.

3 Life in the ice age

In the bleak mid-winter
Frosty wind made moan,
Earth stood hard as iron,
Water like a stone;
Snow had fallen, snow on snow,
Snow on snow,
In the bleak mid-winter,
Long ago.

Christina Rossetti (1830–1894), *Mid-winter*

Point Barrow in northern Alaska is a desolate place. In the driving sleet of an August afternoon it is hard to imagine that this remote Inuit settlement has been occupied for thousands of years. It seems amazing that the community has been able to survive here for so long. The reason is, however, simple. Every year bowhead whales migrate past the point and the Inuit have been able to catch sufficient for their needs and store the flesh and blubber in the permafrost to provide food throughout the year. The reliability of this migratory pattern is the key to continued occupation of the site. The bowhead whale returned to the Arctic Ocean after the end of the last ice age, when the land bridge of Beringia was flooded by rising sea levels around 10 kya. Thereafter, they commonly ranged from the Beaufort Sea to Baffin Bay. Bowheads feed year-round in the Beaufort, Chukchi and Bering seas. They use a variety of strategies, including feeding under ice and swimming in groups in V-shaped formation, to increase feeding efficiency. Their movement into the Beaufort Sea following the receding ice may have been one of the reasons why the Inuit moved into the North American Arctic in the early Holocene. This is but one example of how throughout human prehistory the predictable migration of large animals has been one of the most reliable means of survival. If the climate changed too much and the whale moved away the settlement would have collapsed.

The clear message from all the evidence available on climate change and variability over the last 100 kyr is that the survival of the human race was a precarious business. The ability to interpret the seasonal changes that drove migratory patterns and to plan how to intercept prey was an essential part of survival, especially if combined with a migratory lifestyle of one's own. Until the start of the Holocene around 10 kya, conditions were immeasurably more challenging for humankind. The implications of the challenges of first surviving the more chaotic conditions of the ice age and then exploiting the opportunities that arose with the arrival of the Holocene take us into every aspect of human life.

3.1 THE CLIMATOLOGY OF THE LAST ICE AGE

The previous chapter provided the broad picture of how the climate changed during the last 100 kyr. Our ancestors were, however, concerned only with what the conditions were like where they lived. So it is important to get a better feel of the climatology in different parts of the world, notably where there is archaeological evidence of sustained habitation, and how this changed throughout the ice age. This is a matter of filling in the basic details of the changing regional climate with as much local colour as possible by making the fullest use of the land-based proxy records. In particular, pollen records and beetle assemblages provide insights into the landscape of the time. These sources are patchy, although they are being added to all the time. Inevitably much of the description is linked to specific localities. Nevertheless, it is possible to extend the analysis more widely by taking account of the controlling climatic features at any given time, which were often very different from current conditions.[1]

In relating the evidence of local conditions to the climatic template set out at the end of Chapter 2, there is one basic thing to

[1] General descriptions of the climatic conditions prevailing at different times and place are drawn from a wide variety of climatology texts, the most notable of which are listed in the Bibliography. Where a specific description of conditions at a given place or time relies on a single source it will be identified by a separate footnote.

remember. This is that in the case of the mid-latitudes of Eurasia, much of the evidence of flora and fauna during the ice age will not record the full extent of the dramatic ups and downs that are such a feature of the ice-core records, especially since around 70 kya. Instead, the majority will reflect the most clement conditions and hence gloss over the sudden cold spells that were so much a feature of the glacial epoch. This means that broadly speaking the presentation of regional climates during OIS3 will reflect the interstadial conditions. Conversely, the lengthy period of the LGM provides sufficient evidence to build up an adequate picture of what things were like during the coldest periods. So, inevitably, much of our analysis of the challenges confronting modern humans from the time they entered Eurasia will concentrate on these two extremes.

This approach is in line with the simulations that have been produced as part of the Stage Three Project,[2] which will form the basis of much of what is presented here. This work has produced a comprehensive set of maps of climatic conditions and the biome cover over Europe and western Asia for both interstadial conditions during OIS3 and the LGM. These can be used to explore the conditions during the warmest and coldest conditions during the ice age since around 60 kya. The warm conditions reflect the interstadials identified in the pollen records, and are likely to have coincided with the peaks of the DO oscillations between 60 and 30 kya.[3] As for the conditions during the LGM, although these were probably harsher than during the coldest spells in the preceding 30 kyr, they probably provide a good indication of what things were like during the Heinrich events 2, 3, 4 and 5.

[2] The climatic objectives of the Stage Three Project include identifying the climates and landscapes of Europe during the period 59 to 29 kya and the extent to which climate changes during this period influenced European flora and fauna. Details of the project and its progress can be found on http://www.esc.cam.ac.uk/oistage3/Details/Homepage.html.

[3] Stage Three simulations include attempts to reconstruct conditions for two different types of cold interval, but there was disagreement between the participating scientists as to whether these were accurate representations of conditions at the times in question, as they produce remarkably similar climatic conditions to those computed for the stadials.

Exploiting the work of Stage Three is also consistent with making effective use of the archaeological evidence of modern humans, because there is a distinct bias towards Eurasia, including the Middle East. This emphasis is reinforced by the fact that much of the climatological evidence comes from mid- and high latitudes of the northern hemisphere. This bias does not, however, alter the underlying fact that modern humans were living elsewhere around the world. It is important to inject as much balance as possible by covering conditions in those parts of the world where most modern humans lived for much of the ice age, namely Africa and, to a lesser extent, southern Asia and Australia.

Before embarking upon a geographical and chronological description of the conditions during the ice age, there is one other fundamental feature of the period that must be addressed. This is the influence of the massive ice sheets over North America and Scandinavia. At their greatest extent their influence was huge: not only did they make high latitudes much colder than now, especially in winter, but also their great height meant that weather systems flowed round their fringes rather than penetrating into the Arctic basin. This had the effect of shifting all the weather systems farther south. The full influence of the ice sheets was, however, limited to the LGM. Furthermore, recent studies provide increasing evidence of the ice sheets waxing and waning more than had previously been assumed.

In the case of the Fennoscandian ice sheet, the latest evidence suggests that it did not grow to climatically significant proportions until around 75 kya and remained substantial until about 60 kya. What is now becoming clear is that thereafter it swiftly reduced to small remnants that only briefly expanded and contracted during cooling events (Arnold, van Andel & Valer, 2002). This process continued until around 38 kya, when the ice sheets covered little more than the high mountains of southern Norway and probably parts of the region above 67°–68° N. Subsequently, the climatic signal changed. Although the temperature still fluctuated rapidly, cold events were colder and more long-lasting and allowed sustained ice-sheet growth,

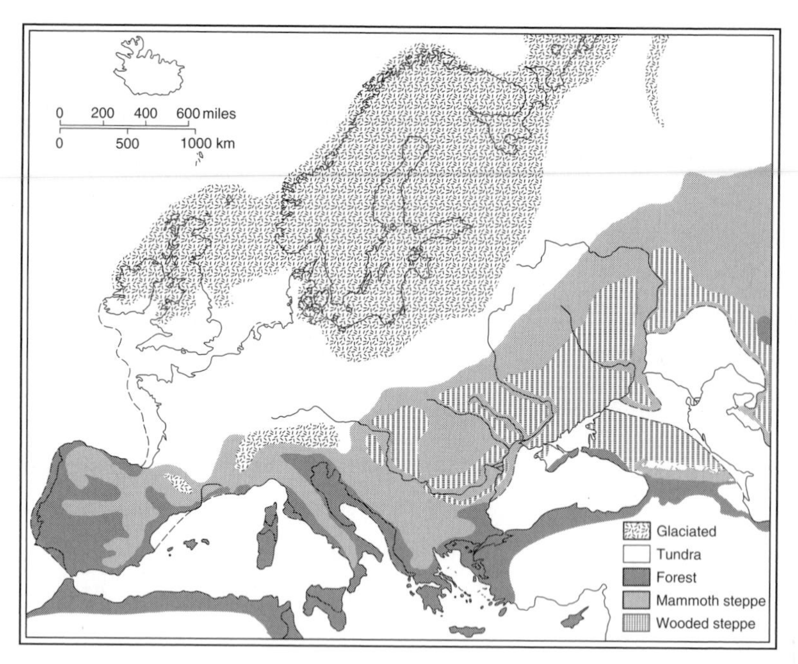

FIGURE 3.1. Map of Europe showing the extent of the Fennoscandian ice sheet during the LGM, and the extent of the continental shelf exposed by the lower sea level at the time.

punctuated by short-lived retreats, until, under a fairly severe climate, the ice sheet reached the Baltic Sea coast about 30 kya. Thereafter, continued cooling enabled the ice sheet to reach its maximum extent during the LGM (see Fig. 3.1).

A similar situation applies for North America. The extent of the ice sheet during much of the last ice age is still a matter of dispute. The sea-level analysis indicates that there must have been a substantial ice sheet over some of the most northern parts of North America. It is less clear how far south it extended and also how big a gap existed between the Cordilleran and Laurentide ice sheets, and until when. What is certain is that during the LGM conditions across the entire continent were dominated by the presence of these two ice sheets (see Fig. 3.2). Indeed the massive scale of these ice sheets is the principal reason we have so little evidence of the extent of earlier ice: all evidence of its

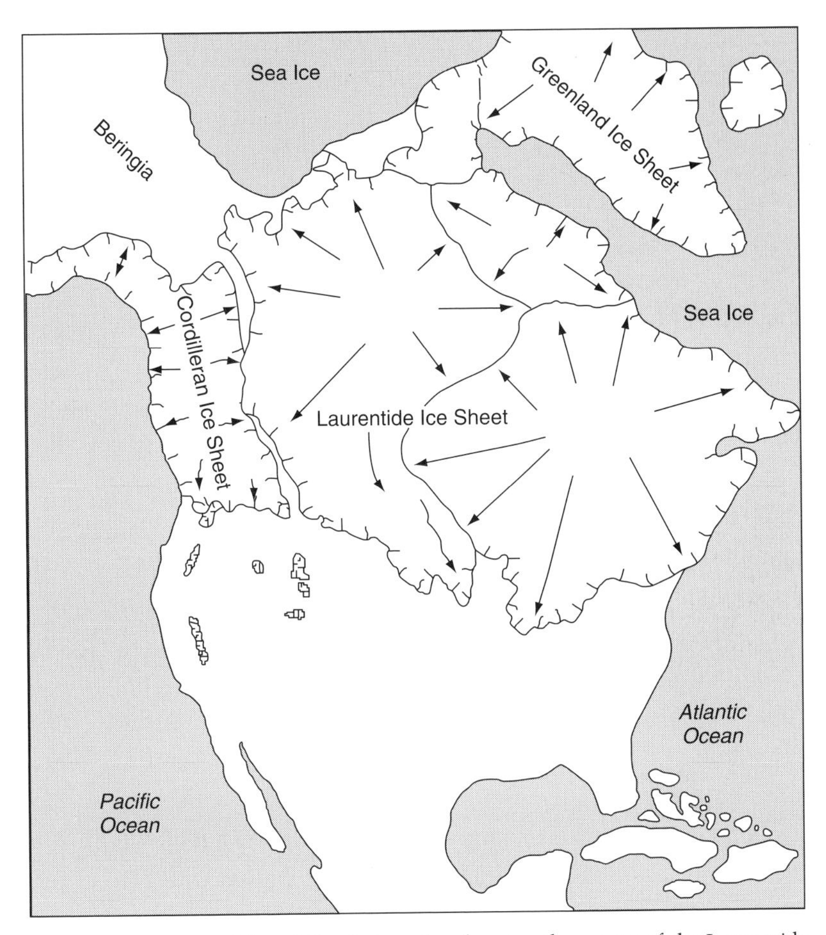

FIGURE 3.2. Map of North America showing the extent of the Laurentide and Cordilleran ice sheets during the LGM, and the extent of the continental shelf exposed by the lower sea level at the time.

past has been scoured from the landscape by the vastness of the final features.

The combined effect of the looming presence of the mighty ice sheets was to push weather patterns and climatic zones farther to the south. In the case of North America this shift led to a squeezing of the circulation patterns. Because the waters of the Gulf of Mexico cooled less than the regions covered by the fringes of the ice sheet, the north–south temperature gradient across what is now the USA was steeper. This differential would have given extra impetus to the storms that generate in the Gulf of Mexico, or sweep in from the North Pacific and across what is now the southern United States.

Across the Atlantic, confronted by the mighty Fennoscandian ice sheet, which would have generated a permanent wintertime anti-cyclone, the storms were more likely to run into the Mediterranean. The evidence is that during the LGM the track was to the south of the Alps and most of the snowfall was on the southern slopes of the mountains (Florineth & Schluchter, 1999). Between the Alps and the ice sheet there was a barren arctic desert. When the Fennoscandian ice sheet was less extensive, the storms moved to the north of the Alps and this transformed the weather in central Europe and the south of Russia and Siberia. So, prior to the LGM, for much of the time modern humans were in Europe this region was habitable and offered an abundant flora and fauna that could be exploited by resourceful people.

The other feature that would have exerted a dramatic impact on the climatology of the northern hemisphere was the extent and seasonal duration of pack ice in subpolar regions. Here again the most important region was the North Atlantic and the controlling factor was the nature of the thermohaline circulation in this region. As explained in Chapter 2, changes in this circulation are closely linked to DO and Heinrich events (Heinrich, 1988; Dansgaard & Oeschger, 1989). The latest analysis of the extent of sea ice is that during the LGM it was more restricted than in the 'classic' CLIMAP reconstruction (Sarnthein, Pflaumann & Weinelt, 2003). During glacial summer, sea ice only covered the Arctic Ocean and western Fram Strait. The northern North Atlantic and Nordic seas were largely ice-free. In winter the ice spread far south across the Iceland Faeroe Ridge, and an extensive patch of sea ice appears to have formed in the central east Atlantic, near the Azores. A broad ice-free channel extended from 50° to 60° N, which sustained the formation of North Atlantic Deep Water (See Section 2.11).

The extreme seasonality in glacial sea-ice formation and melt in the Nordic Seas would have had a major impact on weather patterns.

In summer the high-latitude moisture source would have produced frequent depressions that led to the continued build-up of continental ice sheets. In winter the extensive ice cover would have been an additional factor in driving the storm track farther south and would have reinforced the bitterly cold conditions across northern Europe.

Across Africa the analysis must take on a global tone. While North Africa and Arabia were influenced by patterns that controlled events in the Mediterranean and the Middle East, farther south two other factors came into play. The first was that for much of the ice age the strength of the Intertropical Convergence Zone (ITCZ) was reduced because temperatures in the tropics were lower (CLIMAP, 1981). This meant that rainfall in equatorial regions was reduced and the band of heavy rainfall associated with the ITCZ was restricted to a narrower band close to the Equator. The extent of tropical rain forests was much reduced and savannah was more extensive in equatorial regions. Second, southern Africa was influenced by weather patterns over the southern oceans, rather than events in the northern hemisphere. So to the extent that the southern hemisphere conditions followed a different course from those in the north, climate change in southern Africa was different.

African pollen and lake data indicate that the climate during the LGM was some 4 °C colder and drier than present, with the maximum reduction in precipitation occurring in semi-arid regions. The other thing to note about Africa is that the evidence of changing variability in the climate is missing. Whatever the part played by such shifts in climatic variability in human prehistory in Africa, this is effectively a closed book at present. We know, however, from the studies of the isotope ratios of ancient corals found in Papua New Guinea (see Section 2.7) that the sea surface temperature (SST) in the tropics was lower and the strength of the El Niño / Southern Oscillation (ENSO) was reduced (Tudhope *et al.*, 2001). Because ENSO exerts a dominant influence on many aspects of the variability of the climate in the tropics, it is possible that variability in tropical Africa during much of the last ice age was less than in recent times.

3.2 THE EARLY STAGES OF THE ICE AGE

The fact that the last ice age was not a period of uniform lower temperatures had many implications for humans. The combination of its being much colder than now for much of the time, with the lengthy periods of sustained cooling and warming plus the much greater short-term variability, made life fearfully demanding. If we were to rank the challenges of the ice age, while all substantial changes must have been unwelcome, the sudden coolings probably represented the greatest threat to humans. In particular, those identified as Heinrich events must be put at the top of the list. In all probability, they made large areas of the northern landmasses uninhabitable.

In terms of the challenges to human existence it could be said that the ice age got off to a slow start. The initial stages of the ice age (see Fig. 2.6 and Table 2.1) were relatively limited, but they should not be underestimated. During the last interglacial, which ran from about 130 to 117 kya, global temperatures were for much of the time some 2 °C warmer than current values. Around 124 kya hippopotamus and other tropical fauna roamed the landscape of the British Isles. The latest Greenland ice core (North Greenland Ice Core Project members, 2004) shows the decline in temperature starting around 121 kya and reaching a minimum at 116 kya. Thereafter, the subsequent decline was punctuated by a series of DO events starting around 115 kya and then at 108 and 104 kya.

In terms of the history of modern humans we can pick up the detailed story with the sustained cooling following the interstadial around 104 kya. This cooling continued, interrupted only by a DO event around 91 kya and reached a minimum around 87 kya. In Eastern France during this period the vegetation corresponded to what is now Siberian taiga (Guiot, 1997). This equates to winters being akin to those of modern-day Siberia, somewhere near Irkutz, where the average temperature in January is around –20 °C, some 20 °C colder than current conditions in the Vosges of eastern France.

In the Middle East the most important consequence of the cold conditions around 87 kya was extreme desiccation. The expansion of the deserts here may well have driven out any humans living there at the time. What little evidence there is of climatic conditions in Africa at this time also points to drier conditions. The record of desert-dune formation across southwestern Africa shows that at least part of this period was extremely arid. Desert conditions seem to have existed over a large area west of about 25 °E, and south of about 18 °S. Whether other parts of Africa became arid at about the same time is not known, but given the general pattern in the tropics, this must be considered a strong possibility.

The climate then warmed up around 80 kya with the Odderade interstadial. Then, in terms of sudden climate change, the real action appears to have started around 74 kya. This timing is of particular interest as it coincides with some suggested dates for the movement of modern humans out of Africa. The intensity of the two cooling events at around 74 and 70 kya is greatest in Greenland (Fig. 2.6). Elsewhere the intensity and duration of these events varies. They can, however, be clearly seen in North Atlantic sediment records, speleothems from southwest France to China, and sediment cores from the northwest Indian Ocean (Genty *et al.*, 2003; Wang *et al.*, 2001; Schulz, von Rad & Erlenkeuser, 1998). The importance of these events is that they mark the onset of what appears to have been a more chaotic climatic regime, especially in the North Atlantic. The combination of these two cold events within a few thousand years, both of which were followed by particularly striking DO warming events, stands out as an extreme example of a climate change that could have had a substantial impact on the distribution of modern humans at the time. The scale of both the coolings and subsequent warmings could have posed particular challenges to all forms of life at the time. It is hardly surprising, then, that these events have been singled out for attention.

There is an additional reason for looking closely at the events around 70 to 74 kya. This is the eruption of the supervolcano Toba,

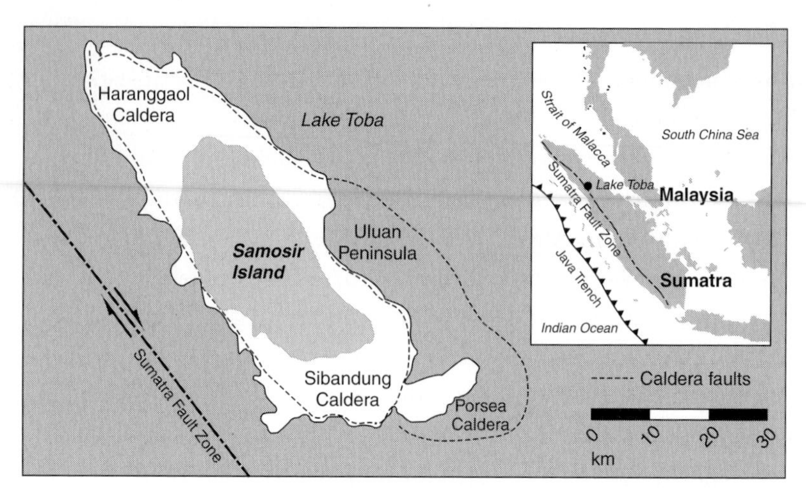

FIGURE 3.3. THE caldera of the 'supervolcano' Toba, which erupted around 74 kya, and which may have led to a dramatic cooling in the global climate at the time.

which is usually dated at about 74 kya (Rampino & Self, 1992). Here is a case where reconciling the various records requires careful detective work. Volcanic ash in sediment cores from the Arabian Sea provides the answer (Schulz, von Rad & Erlenkeuser, 1998). They show clearly that this eruption coincided with the start of the sharp cooling that is seen in the GISP2 ice core at the end of interstadial 20 (i.e. at, say, 71 kya, just before the onset of the second cold event). The discrepancy in the timing of the events is not surprising as the eruption is dated absolutely using radiometric means, whereas the ice core is dated on the basis of modelling the flow characteristics of the ice sheet (see Appendix). What matters is the detailed correlation between the data in the ice core and the ocean-sediment core.

Toba was gigantic. The caldera that resulted from this eruption is 100 km long and 60 km wide (Fig. 3.3). It ejected about 3000 km^3 of material. In the central Indian Ocean, some 2500 km downwind of Toba, a 35-cm-thick layer of ash was deposited. It was the biggest volcanic eruption in the last million years. Furthermore, the eruption was rich in sulphur, which results in the formation of long-lasting sulphuric acid aerosol clouds in the stratosphere. This would have increased its climatic impact; the dust from the eruption would have dropped out of the atmosphere in a matter of months whereas these sulphuric acid aerosols would have remained aloft for several years.

The impact of Toba is estimated to have been about a 5 °C temperature drop, and possibly 15 °C summer cooling in the temperate to high latitudes within a year or so and lasting for several more years (Rampino & Self, 1992). The effects on the growth of plants, and on life in the oceans, of such a dramatic temperature drop would be catastrophic. In many places the dust veil from the volcano would have effectively blotted out the Sun. The cooling would have led to unseasonable frosts in many parts of the world and the disruption of growing seasons. The longer-term impact of Toba is more difficult to establish. It is in the nature of volcanic eruptions that they disrupt the climate temporarily. Furthermore, efforts to demonstrate that Toba was linked directly to subsequent changes in either monsoon rainfall in the Arabian Sea, or temperature trends in Greenland have been less convincing (Schulz *et al.*, 2002).

The dramatic short-term cooling due to Toba may or may not have contributed to the longer-term cooling that occurred subsequently. Either way, such a sudden cooling could easily have had a significant impact on the size of the human population at the time (Ambrose, 1998). Sudden reductions in population are often referred to as 'bottlenecks' (see Section 4.1). Their relevance here is the theoretical possibility that human numbers could have fallen sharply as a result of adverse environmental conditions. As we will see later, this has been postulated as being an explanation of some of the features in the genetic maps of the distribution of different human groups around the world. In this context it is not that important whether Toba had only a temporary climatic impact or precipitated the longer cooling seen in the Greenland ice cores. What mattered was that Toba almost certainly had a dramatic impact on life at the time. Equally certain is that after this cold interval there was a warm period (interstadial 19 in Fig. 2.6).

A more lasting and probably much more damaging cold event started around 67 kya. This is usually linked with Heinrich event 6. It brought periglacial conditions to northern Eurasia. Recent absolutely dated speleothem results from southwestern France show an intensely cold period extending from 67.4 to 61.2 kya (Genty *et al.*, 2003). Here it is estimated that the average temperature dropped by 13.6 °C between 75 kya and 67 kya. The speleothem then stopped growing until 61 kya, presumably because permafrost formed above the cave. These data are supported by the ocean sediment core results (see Fig. 2.7). While ice cores suggest that there was a brief interstadial around 63 kya, for the most part this 6 kyr period was among the coldest during the entire ice age.

3.3 OXYGEN ISOTOPE STAGE THREE (OIS3)

The sustained cold continued until around 59 kya, at which point the conditions started to fluctuate more, and, in particular, there were more periods of relatively warm conditions. This change is identified in the ocean sediment records as being the onset of Oxygen Isotope Stage Three (OIS3, see Fig. 2.6). Although there were strikingly cold periods during OIS3, notably the major cooling associated with Heinrich Event 5 (see Table 2.1), for much of the time much of northern Eurasia was habitable. As a measure of the scale of the fluctuations that occurred during OIS3 the results of the computer model simulations from the Stage Three Project are summarised in Table 3.1. This provides climatic figures typical of the interstadials during OIS3 and the LGM. As noted in Section 3.1, these can be used to provide an indication of the warmest and coldest conditions in both OIS3 and the LGM.

Another aspect of climatic conditions that has been simulated is the duration of snow cover. During the LGM it is estimated that snow lay for the six months of the winter half of the year north of a line from Brittany to the Alps, and east of the Alps to the North of the Danube and then the 50th parallel. In the warmest interstadials this line moved farther north to run from central England across what is now

TABLE 3.1 *A summary of the computer simulations of climatic conditions in various parts of Europe during the Last Glacial Maximum and the interstadials in OIS3*

Region	LGM winter	LGM summer	Interstadial winter	Interstadial summer	Modern values
British Isles	$T \approx -8\,^\circ$C	$T \approx 4\,^\circ$C p: drier than now in lee of ice sheet	$T \approx -4\,^\circ$C	$T \approx 10\,^\circ$C p: comparable with modern values	$T_w \approx 4\,^\circ$C $T_s \approx 15\,^\circ$C
Southwest France	$T \approx 0\,^\circ$C	$T \approx 8\,^\circ$C p: comparable with modern values	$T \approx 2\,^\circ$C	$T \approx 10\,^\circ$C	$T_w \approx 8\,^\circ$C $T_s \approx 16\,^\circ$C
50° N east of 10° E	$T \approx -14\,^\circ$C	$T \approx 12\,^\circ$C (east of 20° E) $T \approx 16\,^\circ$C (east of 35° E)	$T \approx -14\,^\circ$C (east of 25° E)	$T \approx 16$ to $20\,^\circ$C (20° to 35° E) p: drier than modern values	$T_w \approx 0$ to $-4\,^\circ$C (12° to 33° E) $T_s \approx 16$ to $20\,^\circ$C
Northern Balkans	$T \approx -8\,^\circ$C	$T \approx 18\,^\circ$C	$T \approx -4\,^\circ$C	$T \approx 20\,^\circ$C (up to 50° N east of Black Sea)	$T_w \approx 0\,^\circ$C $T_s \approx 22\,^\circ$C
Southern Italy	$T \approx -2\,^\circ$C	$T \approx 16\,^\circ$C p: wetter than modern values	$T \approx 0\,^\circ$C	$T \approx 18$ to $20\,^\circ$C p: drier than modern values	$T_w \approx 6\,^\circ$C $T_s \approx 24\,^\circ$C

Details of this project and its progress can be found on www.esc.cam.ac.uk/oistage3/Details/Homepage.html.
p: precipitation.

the North Sea and approximately along the 55th parallel. The corresponding line for three months of winter snow in the LGM ran from southern Brittany to the South of France across central Italy and the southern Balkans to the Black Sea. During the interstadials this line ran from southern Ireland to the Alps, then along the Danube Valley to the northern shores of the Black Sea.

What the Stage Three simulations cannot yet handle is the shorter-term fluctuations in the record. Attempts to simulate conditions during the frequent stadials in OIS3 produced results that the various groups working in the field could not accept and further work is in progress. These difficulties may arise from the fact that the computer models cannot reproduce the dramatic short-term variability of the ice-age climate (see Fig. 2.9). This variability may also explain the fact that the pollen records do not appear to reflect all the shorter warm episodes seen in the ice-core data. These warmer conditions show up as the spread of trees northwards, and this would have taken some time to react to the sudden shifts associated with DO oscillations. In addition, the extreme interdecadal climatic variability probably acted as a severe brake on recolonisation.

It is not surprising that the computer models have difficulty in handling the ups and downs during OIS3, nor that the various proxy records are hard to reconcile. Nevertheless, these records are producing an increasingly detailed and consistent picture. Ice-core and ocean-sediment records plus results from speleothems and lake-sediment cores enable us to date climatic events across the continents of the northern hemisphere (see Appendix). These measurements also provide greater insight into changes in weather patterns at the time and can add weight to conclusions drawn from analyses of pollen records and beetle assemblages. For instance, oxygen isotope records in five stalagmites from Hulu Cave near Nanjing, in China, covering the period 75 to 11 kya, which provide a measure of the monsoon rains in the region, bear a remarkable resemblance to oxygen isotope records from ice cores (Wang *et al.*, 2001). This supports the idea that millennial-scale fluctuations identified in Greenland are hemispheric or even wider in extent. It suggests that the circulation of the northern hemisphere was more meridional during Greenland interstadials and more zonal during stadials. So, changes in the East Asian monsoon are an integral part of these circulation patterns.

In terms of human prehistory, what makes OIS3 of particular interest is that after Heinrich event 5 there was a relatively mild period that appears to have coincided with the Upper Palaeolithic Revolution (see Section 4.3). Using the figures in Table 3.1, it is reasonable to assume that around 40 kya, during the marked interstadial, the conditions across the North European Plain the mid-summer temperatures varied from about 15 °C in western France to over 20 °C in southern Russia: not much cooler than at present. The vegetation was largely herbaceous, with a rich variety of species that adapted to local conditions. The absence of birch or pine/spruce does, however, support the idea that short-term variability hindered the growth of longer-lived species. Furthermore, during the markedly colder winters there was snow cover for across most of the North European Plain for three to six months. This, combined with stronger winds, rapid spring thaws and flooding, and highly unstable soils made the conditions more like those currently experienced on the river floodplains of Siberia. The open habitat may also have been the result of high grazing intensity by large herbivorous animals such as woolly mammoths, woolly rhinoceroses, reindeer and bison.

Before and after this temperate interlude there were phases of colder, more continental climate during which summers were too cold for trees to grow and winters were of even greater intensity. Just how cold the worst periods were is difficult to judge, as the intense cold so reduced growth that no fossil evidence can be found in northern Europe. The only evidence that can be found for these coldest periods is in the fossil records from the warmer parts of southern Europe and the information in ocean sediments and the Greenland ice cores.

Across the Mediterranean and into the Middle East there is plenty of evidence that events in the North Atlantic exerted a considerable influence on climatic conditions. Possibly the most important aspect of these changes in terms of human activities was the changes in precipitation regimes, as temperature ranges were hospitable compared with farther north. In Israel a detailed lake-level history of the closed Lake Lisan (part of the ancient Dead Sea) indicates that for much of the period from 55 kya to the end of the ice age the level was higher than at present (Bartov *et al.*, 2003). There were, however, catastrophic droughts associated with the Heinrich events. The impression is that cold-water input to the Mediterranean originating in the collapse of North Atlantic Deep Water formation caused a reduction of evaporation and less precipitation in the Levant. So it is reasonable to conclude that following Heinrich events 6, 5, 4 and 3 (see Table 2.1) much of the Middle East was desert. During these intervals, of which the long cold period from around 67 to 59 kya was the most significant, the region was largely uninhabitable for modern humans.

The evidence of the striking parallel between events in Greenland and the North Atlantic extends into the northwestern Indian Ocean (Schulz, von Rad & Erlenkeuser, 1998). Sediment cores from this region show a close correlation with DO and Heinrich events, suggesting that these North Atlantic events were a significant component of low-latitude climate change during the last ice age. These millennial and centennial fluctuations would have been reflected in the strength of the southwest monsoonal circulation over south Asia.

Across Siberia the pattern of changes was similar to that in northern Europe. The warmer conditions that occurred during OIS3 were much like what would be expected today during the summer half of the year. Evidence of interstadials appears in the pollen records in southern part of Siberia during the periods 43–34.5 kya, possibly associated with the warmest parts of some of the DO events 5 to 12. Reasonably large areas of taiga vegetation appear to have existed in the Central Asian Altai Mountains during at least parts of this period. The warmth was such that it enabled humans to occupy parts of

Siberia as far north as the Arctic Circle on the banks of the Usa River, close to the Ural Mountains, nearly 40 kya, and this occupation appears to have extended to the Yana River valley within 2000 km of the Bering Straits by 30 kya (Pavlov, Svendson & Indrelid, 2001; Pitulko *et al.*, 2004).

There is a more general feature of the conditions across Siberia during OIS3. This is the high productivity of the region in terms of fauna. The vast herds of herbivores appear difficult to reconcile with current tundra conditions in northern Siberia. This has led to the concept of the 'mammoth steppe' as being a region where the combination of low precipitation, and hence light snow cover, plus abundant sunshine in summer led to a biome that is more reminiscent of the prairies of the northern US and southern Canada (Guthrie, 1990). This hypothesis is supported by recent studies of ice-age steppe vegetation in the Yukon Territory, Canada (Zazula *et al.*, 2003). The core of the mammoth steppe region was central Asia but in the warmer periods of the warmer parts of OIS3 it extended into western Europe and right across Beringia. As we will see, this more productive image of this steppe has important implications for the spread of modern humans across Eurasia and into the Americas.

Another reason why Siberia and Beringia may have been more habitable for humans was that, according to the evidence, the extremes of climate change in the North Atlantic became more muted farther east across Eurasia. This can be seen most clearly in Beringia. As with much of Siberia, there was a maximum in forest development across Beringia between around 33 and 39 kya. In general, in the east of the region vegetation showed less propensity to swing between stadial and interstadial conditions, whereas near-interglacial forests alternated with more glacial-like tundra in western Beringia. The 'flickering' of the interstadial forests suggests great climatic variability in the west, in contrast to the more stable climatic regime farther east.

There is less information about conditions in North America during OIS3, especially across the Great Plains. Studies of stalagmites from a cave in southeastern Missouri show remarkably little variation after a marked cooling around 55 kya (Dorale *et al.*, 1998). Prior to this, conditions appear to have oscillated more frequently. The warmest temperatures occurred around 57 kya and there were short-lived cooling events around 64, 71 and 74 kya. Beetle assemblages suggest that, following the cooling around 55 kya, the mean July temperature was 7.5–8 °C lower than present and mean January value was 15–18 °C lower than present. During what is termed the 'mid-Wisconsin' interstadial, dating from 43.5 to 39 kya, there was a rapid and intense warming (Elias, 1999). At the peak of this event, about 42 kya, July temperatures were only 1–2 °C lower than modern. Farther south and east pollen analysis of lacustrine deposits in Lake Tulane, in northern Florida, provides a more detailed picture. Here, the movement of vegetation zones back to 50 kya tallies well with events in the North Atlantic (Grimm *et al.*, 1993).

The big gap in the North American data is the lack of knowledge of fluctuations in the extent of the Laurentide and Cordilleran ice sheets during OIS3. In part, this is an inevitable consequence of the fact that the subsequent expansion of these ice sheets erased all evidence of the previous build-up. There is a general assumption that they were considerably less extensive between 60 and 30 kya. What is less certain is whether they receded as much as the Fennoscandian ice sheet during the warmest times during this period. On the basis of sea level measurements we have to assume that there were considerable ice sheets somewhere in the northern hemisphere at the time. So, if they were not over northern Europe, they had to be over eastern Canada and possibly the western Cordillera. The intriguing unanswered question is how accessible North America was from Beringia, as this is part of the puzzle about the peopling of Americas (see Section 5.12).

Elsewhere, a similar pattern of changes occurred. Around the time of the onset of OIS3 the Australian climate entered a wetter phase. These moister conditions prevailed until around 40 kya, with only occasional drier periods lasting up to a millennium, although there is evidence of an increasing incidence of dry periods towards the end of this period (Bowler *et al.*, 2003). Then the climate became much more arid. The transition to increasing aridity appears to have coincided with the arrival of humans in Australia, which is now estimated to have been around 60 kya (see Section 3.8).

3.4 THE LAST GLACIAL MAXIMUM (LGM)

The harsh conditions during the LGM (see Table 3.1) epitomise the ice age at its worst. The end of the Heinrich event 3 at around 29 kya marked the termination of OIS3. In the ocean-sediment records the next stage (OIS2) extends to around 15 kya (see Fig. 2.6). Although this demarcation coincides with the rapid fall in sea level and the expansion of the Fennoscandian and Laurentide ice sheets, in practice there were some relatively warm spells between 30 and 25 kya, notably in southern Siberia. Here a warmer episode is marked by a predominance of tree pollen dated to around 30–25 kya, apparently associated in some way with the DO events 2 and 3.

The true LGM falls in between the Heinrich events 2 and 1 at around 23 kya and 16.5 kya. Its impact was particularly severe on northwestern Europe. The massive extent of the Fennoscandian ice sheet (Fig. 3.1) reduced much of the region north of the Alps and the Pyrenees to a polar desert. Permafrost extended down to southern France, just north of Bordeaux, and into the uplands of northern Provence. The areas just to the south of the main ice sheets had little or no vegetation. Dune activity during the LGM seems to have been quite widespread in England, northern France and the Low Countries, and eastwards across Germany and Poland. In effect, for humans these areas were uninhabitable. Only in the Dordogne region of southwest France and the foothills of the Pyrenees, where modern humans had lived for many millennia before the LGM, is there widespread evidence of their having stuck it out during the LGM.

Even in the Mediterranean region, there appears to have been little thick woody vegetation. Here the predominant picture was of an arid semi-desert. The occasional small pockets of open woodland where local soil moisture levels permitted would have broken this desolate landscape. The refugia for deciduous and needle-leaved species were mainly on the western side of the mountains of Greece. Paradoxically, lake levels were high, possibly as a consequence of relatively high winter rainfall, in the form of intense storms with high run-off. This had the effect of filling the lakes without making much moisture available to the plants. Lower year-round temperatures would have suppressed evaporation from these lakes.

In central and eastern Europe it might be assumed that the conditions were even less hospitable for humans, but not so. The landscape was desolate with a few cold-tolerant trees (pine, birch and spruce, for example) in isolated pockets. Further east, in the Russian steppes, woodland seems to have survived along river valleys. Elsewhere, it was predominantly open steppe or steppe-tundra. In many places, where there was inadequate vegetation, windblown dust formed dunes. Precipitation, now around 600 mm, may have been only about 60–120 mm per year. In spite of all this, humans do seem to have been present in some places for much of the LGM. A number of well-dated encampment sites have been found along the valleys of both the Danube and the Don. Furthermore, some refuge areas of woodland may have existed immediately to the southwest of the Carpathians.

Along the eastern shores of the Black Sea and the high ranges of the Caucuses, pollen evidence indicates that there were scattered pine and birch forests but broad-leaved trees were probably localised in distribution. Only small areas of dense temperate forest appear to have survived in the lowlands of the southern Caucasus. These are thought to have been a glacial refuge for many temperate trees, although there is little direct evidence of these trees surviving there during the LGM. The Black Sea was shallower and smaller than today, while the Caspian Sea was somewhat deeper and more extensive. There may have been a thin band of deciduous forest along the southern shores of both these seas.

Across Asia Minor the evidence suggests much less woodland and more steppe and semi-desert steppe than now in the upland areas of Turkey, northern Syria and western Iran. Open woodland or wooded-steppe may have survived over much of western, southern and eastern Turkey. Woodland is thought likely to have been present along the western Levant. As noted in Section 3.3, evidence from the Dead Sea (Bartov *et al.*, 2003) suggests that, at the beginning and end of the LGM, catastrophic droughts in the Eastern Mediterranean coincided with Heinrich events 1 and 2. In this context, it is notable that the settlement at Ohalo II (see Section 3.13) appears to have been abandoned at about the time of the start of event 2.

Farther east, the general situation across Siberia was that the climatic zones were pushed to the south. The absence of a polar ice sheet east of the Urals meant that in some respects the consequences of the LGM were less exaggerated here. The evidence suggests that winter temperatures across southern Siberia were about 12 °C lower than now being comparable with those in northeastern Siberia at present. Summer temperatures are reckoned as being about 6 °C lower throughout Siberia and the central Asian desert region.

The other feature was that to the south, it seems, there was not a large ice sheet over the Himalayan Plateau, but rather a scattering of glaciers and small ice caps. Permafrost desert conditions appear to have existed in the unglaciated parts of the mountains. A consequence of this lack of ice cover and the general aridity is that it may have made parts of southern Siberia to the east of Lake Baikal more congenial than might have been expected. For instance, to the west of Lake Baikal at Mal'ta around 23 kya the vegetation was steppe merging into tundra with lakes, small streams and rivers that were crisscrossed by reindeer migratory routes and rich with waterfowl. The relatively less extreme climate is probably the reason why this part of the world seems to have remained habitable throughout the LGM.

Even more interesting is the question of whether Beringia effectively represented a refuge (see Section 3.12) during the LGM. Analysis of beetle assemblages provides interesting insights into the climate during the ice age (Orlova *et al.*, 2001). At times temperatures in the region were relatively high. The Stage Three evidence (see Section 3.3) suggested that at times northeast Siberia and Alaska were surprisingly mild. During the interstadials of Stage Three the region had temperatures that appear to have been comparable with modern times. Even as late as 30 to 25 kya parts of northeastern Siberia experienced summertime temperatures close to modern values. This relative warmth appears to have continued into the LGM. At the time, unlike the North Atlantic, the northern Pacific and the Gulf of Alaska were largely free of sea ice. This would have led to maritime cloud cover spreading over the extensive plains between what is now Chukotka and Alaska. The climate would have been colder than now in summer, but relatively mild in winter. Combined with the extensive megafauna of the region, this may have made parts of the region habitable during the LGM.

Moving to southern Asia, the problem with putting the climatic conditions during the LGM into a human context is that, in this part of the world, there is relatively little evidence about human activities at this time. The overall picture is one of aridity. This suggests a failure of the summer monsoon rains to penetrate as far north as at present. Information is, to say the least, spotty. The best information comes from China, and paints a general picture of a major reduction in forest vegetation and southwards retreat of climatic zones relative to the present.

Considerable stretches of low-lying land were uncovered around the shores of China, and the Malaysian Peninsula became linked to the islands of Borneo, Java and Sumatra, and to the Philippines. The islands of Japan were linked together into a peninsula owing to the lower sea level, but probably remained separated from the Asian mainland by the Korean channel. The Sea of Japan was almost entirely enclosed as a lake, its only outlet being the Korean channel. So there was the scope for people to move around these regions, but more important, where they lived close to the sea, all evidence of their activities has been lost.

Northern China was much colder and more arid conditions prevailed than at present. The summer monsoon limit was shifted some 700 km to the southeast. The widespread distribution of loess indicates more extensive central Asian desert conditions with low biological activity and a sparse herbaceous vegetation cover. Fossil evidence suggests that the mammoth disappeared during the LGM, whereas they were abundant before and after this period. The upper limits of trees in mountainous areas in both northern and southern China were some 1700 m lower than at present.

Farther south in east China and Taiwan, which was connected to the Chinese mainland during the LGM, dry steppe vegetation, with some pine trees in a wooded steppe, covered much of the lowlands. On what are now the highest rainfall areas in the uplands of Taiwan, pollen evidence indicates some forest vegetation persisted. Scattered areas of wooded vegetation covered about a third of the region. In southernmost China the climate was much closer to current conditions, and hence well suited to human habitation. Nevertheless, the subtropical rainforest was replaced by mixed conifer and evergreen broad-leaved forest. Grasslands predominated in lowland areas, with cool temperate forest and open woodlands in upland areas. In the mountainous areas of Northern Yunnan Province of southwest China, there are indications of snowline lowering, indicating a 4–5 °C depression in temperature and slightly moister conditions than now. In the present subtropical rainforest zone in the uplands of Yunnan Province the climate was nearly warm as at present, but with much higher precipitation in winter.

In Japan, where lower sea level provided land links between the four major islands and the Asian mainland, there is evidence of modern human occupation from around 30 kya. Here, there was a southwards shift of the vegetation zones. Permanent ice seems to have covered the uplands of what is now Hokkaido, with a belt of tundra and open boreal woodland in the lowlands. Farther south the lowland grassland with scattered stands of alder, ash and willow. Forests of a rather open character with oak and pine seem to have been widespread in the mid-altitude uplands. Open woodland, consisting mainly of pine and birch, covered much of Japan's uplands, from about the middle of the main island to the south of the linked chain of islands. Here, trees that prefer warmer temperate conditions, such as the cryptomeria, persisted only locally in the lowlands of southern Japan. Mean annual temperature here seems to have been about 7–9 °C lower and precipitation was probably less than a third of present values.

There is little information from Indo-China and Malaysia. There are indications of pine forest occurring in the present rainforest areas of Thailand and Malaysia. In Sumatra and west Java the climate appears to have been drier. There also seems to have been a lowering of the mean annual temperature by about 4–7 °C. In the highlands of these islands and New Guinea, where modern humans have lived since 40 kya, it was 2–3 °C cooler, but not drier. In what is at present an extremely wet rainforest climate (3200–5000 mm annual rainfall) in lowland western Borneo (Kalimantan), there is evidence of savannah development. In the present rainforest region of Sarawak and Sabah (northern Borneo), the rainforest persisted through the LGM. Other evidence suggests that there would have been an arid climate and sparser vegetation over most of the exposed the exposed continental shelf between the islands of Indonesia that is often termed 'Sundaland'.

India seems to have been much drier and more sparsely vegetated at the LGM. In the northwest, there were fairly widespread desert conditions. Salinity in the northern Arabian Sea appears to have been higher than today, indicating decreased input from rivers. This suggests that reduced rainfall and reduced run-off of rivers from the Western Ghats. Southern India was also more arid than at present. With lower sea level, Sri Lanka would have been connected to mainland India and the exposed Palk Strait may have been covered in dry forest or savannah-like vegetation .

Across North America the shifts in climate patterns were dominated by the massive presence of the huge ice sheets that covered virtually all of Canada, the upper mid-West and New England. This meant that all the climatic zones were squeezed southwards. As the tropical oceans cooled less, however, the Gulf of Mexico remained relatively warm and so the winter temperature gradient across what is now the USA was steeper. So conditions that are now the norm from southern New England to northern New Mexico were more typical of northern Florida and along the coast of the Gulf of Mexico. Farther west the shift southwards was somewhat reduced. The impact on summer temperature patterns was comparable. Beetle assemblages show that during the Heinrich 2 event the July temperature in southern Missouri was 10–12 °C lower than modern values (Elias, 1999).

3.5 THE IMPLICATIONS OF GREATER CLIMATIC VARIABILITY

The huge climatic changes that occurred during the last ice age are only part of the story. Any analysis of the conditions prior to 12 kya must take full account of the greater variability of the climate (see Fig. 2.9). Those of us living in the mid-latitudes of the northern hemisphere tend to think that we live in pretty exciting times when it comes to variable weather. The fluctuations from week to week or from winter to winter in Chicago or Stockholm are the source of continual discussion. So it is hard to imagine what it might be like to experience the extreme variability of the last ice age. This was not simply a matter of where humans lived being much colder than now. Around 20 kya the farthest north people were living in Europe was probably around the Pyrenees in France and Cantabria, in northern Italy or the northern Balkans and parts of southern Russia. Here the average conditions might have been no worse than currently experienced in southern Alaska, central Russia or Siberia. The fluctuations from year to year would, however, have added an extra dimension to the challenge of living.

Available records for western Europe and the eastern United States,[4] covering the last two to three centuries, provide us with a

[4] There are a number of lengthy temperature records for various places in northwestern Europe. Of particular value in the current analysis are the Central England Temperature record (Manley, 1974), which provides monthly figures back to 1659, and the composite record built for de Bilt in the Netherlands (van Engelen & Nellestijn, 1995). The variance figures quoted here represent an average of these records.

measure of the current variability of our climate. On a seasonal basis the difference between the coldest and warmest winter since 1700 in western Europe has been about 10 °C. At one extreme, much of the continent would be covered by deep snow and most rivers and large lakes frozen solid for much of the time between early December to early March; at the other extreme is a largely frost-free winter with many trees and shrubs coming into bloom in January. The greater variability prior to 12 kya suggests that changes of this order would have occurred within a few years of one another as a natural part of these more variable conditions.

In summer the differences in our modern climate are smaller, with the temperature range between the warmest and coolest seasons in the last 300 years being just under 5 °C. Nevertheless, when combined with radically different rainfall patterns and amounts of sunshine, the impact on agriculture has been dramatic. Wine harvests ranged from early and abundant to late and pitiful with harvest dates varying by over 50 days between the earliest and latest. Grain harvests showed a more complicated response, with average conditions producing the best results and extreme heat and drought, or exceptional wet, cool and cloudy conditions, producing dramatic falls in production. In the Middle Ages the most damaging conditions were cool, wet cloudy weather throughout the growing season that delayed the harvest into autumn and made the gathering of crops almost impossible. In the worst summers, such as 1316, when in the fertile lands of the Bishopric of Winchester, in southern England, only two grains were reaped for every one sowed, famine stalked the land. As we will see, this is too low to sustain agriculture and well below the average figures achieved from the earliest days of agriculture. Clearly, if extremes like this were to become more frequent they would pose awful challenges for agriculture.

In the eastern half of the United States the range between extreme winters since the late eighteenth century is comparable to the figure of 10 °C quoted above for northwestern Europe. Based on the experience of the twentieth century, these fluctuations can lead to a similar scale of social disruption. In summer, the fluctuation in mean seasonal temperature is smaller, but in terms of agricultural impact it is the variations in rainfall that matter most, with the hot dry years causing the worst damage, as witness the events of the 'Dust Bowl years' of 1934 and 1936. The fluctuations in the levels of dust, calcium and sea salt in the Greenland ice cores (Taylor *et al.*, 1993), together with measurements of annual snowfall, suggest that the variance in the strength of circulation and hence mid-latitude precipitation levels was correspondingly greater before the Holocene.

As for longer-term changes in the scale of the variations in recent centuries, these have also been comparatively minor. From the coldest to the warmest years the range is between 3 and 4 °C, while between the warmest and coldest decades the range is between 1 and 2 °C. Longer-term trends in western Europe amount to about 1 °C over the last three centuries in wintertime, and somewhat less in summertime temperatures.

The figures that emerge from the ice-core records for the time before 12 kya are starkly different. While we have to be careful about making comparisons between isotope data for Greenland and events in, say, Europe, we are talking about substantial differences. As was noted in Chapter 2, the suddenness of the major changes in climate was unlike anything we have experienced in recorded history. In Greenland the difference between the coldest parts of the last ice age and current conditions was some 20 °C. The sudden jumps in temperature associated with Heinrich and DO events were typically in the range 5 to 10 °C. Furthermore, sustained trends over several centuries produced comparable shifts. But in terms the impact of variability on human activities it is interdecadal variance that mattered most. Effectively the interdecadal variance and its impact were some tenfold greater (see Section 2.7). These wild swings must have been immeasurably more demanding than our present climate. They would have required an extraordinarily adaptable, flexible and migratory lifestyle to adjust to changing environmental conditions. At the simplest level, it is probably true to say that even now such a climate would make any form of agriculture, as we currently know it, virtually impossible: an issue we will return to later.

Alongside placing immense demands on the resourcefulness of the communities that survived in such a climate, this variability would have placed a particular premium on exploiting the most 'favourable' environments. The word favourable is set in inverted commas, as it requires definition. Climatic favourability might encompass being relatively close to the sea to reduce temperature fluctuations, or it could involve shelter from prevailing winds together with having soil conditions extending over a considerable altitude range that enabled vegetation to adapt to climate change locally. The existence of such 'refugia' is discussed in Section 3.12.

3.6 LOWER SEA LEVELS

In terms of living conditions, the lower sea levels (see Chapter 2) were a significant benefit for modern humans. Around the world they exposed up to 25 million square kilometres of continental shelf. This had a number of important implications for the migration and survival of the human race. It also has profound consequences for our reading of the archaeological record, as much of the evidence of life during the ice age has either been swept away or lies buried below sediment under the sea.

At low latitudes, perhaps the most important feature of the exposed continental shelf was that it made it easier for people to move about. In particular, in the Persian Gulf, around India, and, most of all, down through southeast Asia and Indonesia the linking of many of the islands (Sundaland) greatly assisted human mobility. The drop after 85 kya, and the low level between about 67 and 61 kya may have played a crucial part in the movement out of Africa and the early arrival of modern humans in Australia, although they still had to overcome the considerable challenge of sailing across the much reduced Timor Sea (Fig. 3.4). The same applies to the land bridge that formed between northeastern Asia and Alaska (Beringia), which is

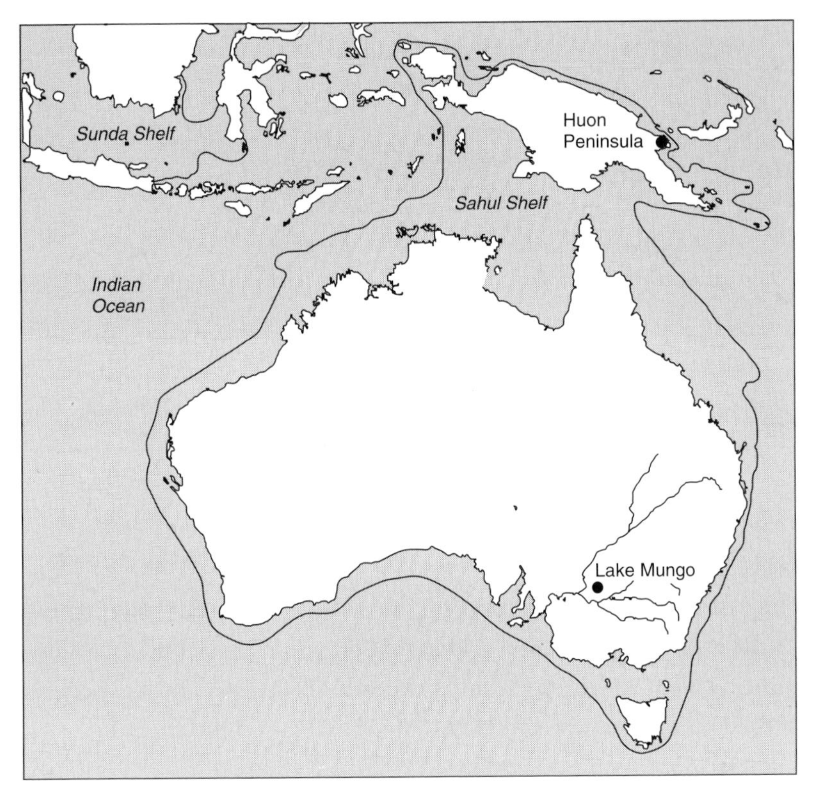

FIGURE 3.4. Map of Australia and Papua New Guinea during the LGM, and the extent of the continental shelf exposed by the lower sea level at the time. Sites mentioned in the text are marked.

regarded as the only feasible route for modern humans to reach North America (Fig. 3.5).

The other essential feature of the lower sea level during the ice age was that it exposed large areas of attractive living space. It is likely that many of our forebears exploited this space, not least because the seashore offered an abundant source of food, notably in the form of shellfish. Virtually all evidence of this occupation has been lost beneath the rising sea. Furthermore, this inundation has been followed by sediment forming to cover up what traces remain of human activities. Only surveys with modern sophisticated imaging techniques are likely to enable us to find out a little bit more about

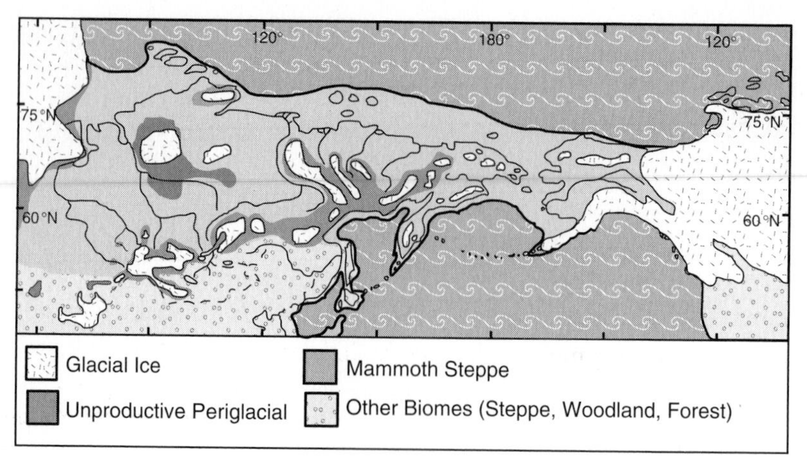

FIGURE 3.5. Map of eastern Siberia, Asia and western North America showing the region covered by Beringia, together with the patterns of vegetation and the extent of the ice sheets during the LGM.

this lost world. Such studies could throw an entirely new light on life on the exposed continental shelves of Europe and Asia, but also examine the extent to which the areas exposed off the east and west coasts of the United States were ever occupied by Native Americans.

3.7 GENETIC MAPPING

Linking of climatic conditions in different parts of the world with human prehistory requires us to find out where people actually lived during the ice age. Before we can start looking at how the archaeological record can be interpreted in terms of an up-to-date template of climatic change, we must introduce another great scientific advance: genetic mapping.[5] This analysis of the common origins of modern humans and how they spread out across the globe is central to any understanding of the extent to which human survival and social development was governed by climate change. At least 20 groups worldwide are engaged in studies of genetic variation to find evidence of our past. Sometimes called molecular anthropologists, these

[5] The presentation here draws heavily on the various books cited in the Bibliography that provide an introduction to the rapid development in genetic mapping in recent years.

researchers analyse DNA to identify *genetic polymorphisms* (situations where genes exist in different forms in different people's DNA). They use these markers to interpret human evolution and population migrations.

The principle underlying the mapping of genetic markers is that the difference in the frequency of the form of a gene in different populations is a measure of the time since these populations separated. This difference in frequency is usually termed the *genetic distance*, and is best measured using the largest number of genes that is practically possible. Generally, the genetic distance increases with geographical distance, although physical barriers such as seas and mountain barriers complicate this measure.

Two developments have boosted the endeavours of molecular anthropologists to map human prehistory. Scientists have discovered enough useful DNA markers over the past 20 years to develop a structured approach to their analysis. In addition, technology has come on in leaps and bounds. Nevertheless, these developments could not tackle the challenge of looking at all our genes. The way to simplify the analysis is to focus on those genes that are *haplotypes*. These occur in chromosomes or sequences of DNA that are *haploid* rather than *diploid*. Haploid chromosomes contain genetic information from only one parent, whereas diploid chromosomes contain information provided from both the mother and the father in each generation. Haplotypes consist of several genes or DNA variants inherited as a linear cluster. Those in mitochondrial DNA (mtDNA) are inherited through the female lineage, while those on the sex-specific portion of the Y-chromosome pass down through the male lineage.

Initially, the principal effort was devoted to tracing maternal ancestry through mtDNA. Human eggs hold some 100 000, mitochondria, each with a circular mtDNA of 16 569, nucleotides. This DNA is passed exclusively from mother to child. Differences between mtDNA sequences are only due to mutation. Any mutation is passed on from one generation to the next. So, as time passes, mutations accumulate sequentially along hereditary lines. Using what are called *phylogenetic networks* or *trees*, in which mutations are classified in hierarchical levels, it is possible to estimate relationships among lineages. Basal mutations are shared for clusters of lineages, defined as *haplogroups*, whereas those at the tips characterize individuals.

The logic of this construct is that at the root of the phylogenetic network is a single woman. This analysis led to the 1987 proposal by Allan Wilson at the University of California, Berkeley, that a 'mitochondrial Eve', who lived roughly 200 kya, was the source of all people's mtDNA (Cann, Stoneking & Wilson, 1987; Wilson & Cann, 1992).[6] Since that groundbreaking work, various proposals have been made for grouping humans into a number of lineages based on mtDNA markers that are the surviving lines from the earlier lineages that extend back to the original Eve.

Major haplogroups are continental or ethnically specific (Fig. 3.6). Three of them (L1, L2 and L3) are sub-Saharan African lineages; nine (H, I, J, K, T, U, V, W and X) encompass almost all mtDNAs from European, North African and Western Asian Caucasians. The remaining haplogroups A, B, C, D, E, F, G and M embrace the majority of the lineages described for Asia, Oceania and Native Americans. The geographic distribution of derived branches of these haplogroups has shed light on crucial aspects of human history, such as the probable origin and approximate dating of migrations into the New World and Polynesia, and quantitative estimations of the relative proportions of Palaeolithic and Neolithic populations in Europe. The chronology of the branching of these lineages has not only produced some fascinating insights into human history, but is still a bone of contention between some palaeontologists and molecular anthropologists. This combination of cutting-edge science and professional rivalry is the stuff of some excellent popular science books (see Bibliography).

[6] The term 'mitochondrial Eve' is in some ways intensely misleading, as it can give the impression that there was a single woman living at the time in some African Garden of Eden. In practice, there would have been a considerable number of other women living at the time. Eve's pre-eminence is that only her mitochondrial line has survived to feature in all living humans: all the other lines died out long ago.

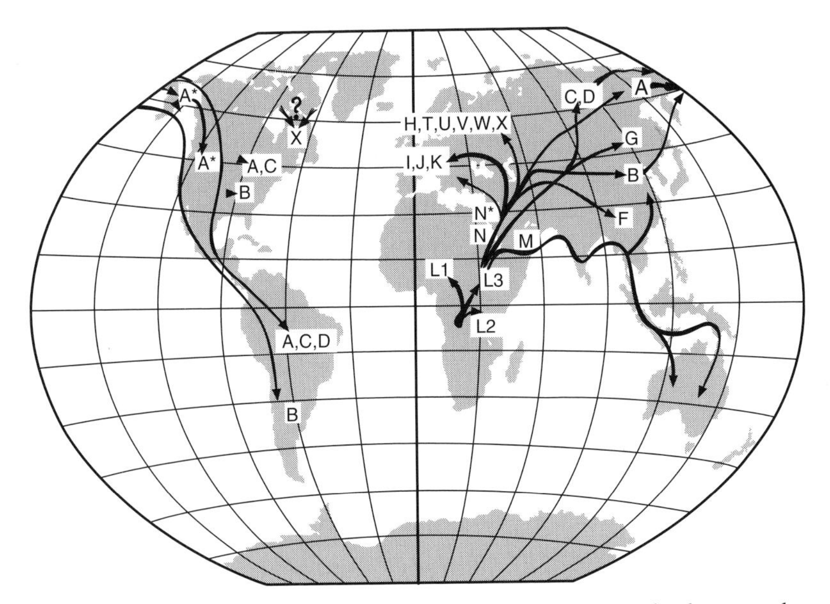

FIGURE 3.6. A map of the global distribution of the main haplogroups that have been used to identify the migration of modern humans out of Africa. This figure should be read in conjunction with Fig. 1.4, which shows the approximate timing of these movements across the globe.

Much of the debate about the reliability of mtDNA studies rests on two factors. One is the question of whether relying on the special properties of the control region may disguise more complicated features of the population dynamics of modern human development. The other is that concentrating on the female side of the equation fails to capture the possible consequences of the fact that some males may father large numbers of children. As noted above, men's genetic history is recorded on the sex-specific portion of the Y chromosome. By comparing polymorphisms at this site on the chromosome, and the frequency at which those variations occur in different populations, geneticists can sort out which populations are most closely related. They can then build a phylogenetic tree that traces the inheritance of the Y-chromosome markers in different populations. Then, just as with mtDNA studies, by using average mutation rates, researchers can estimate how long ago particular mutations appeared, thus dating various population splits and movements.

Until several years ago, few Y-chromosome markers were known. This changed in the mid-1990s. Now it has been possible to trace human lineages back to what is termed '10 sons of a genetic Adam'. As with the lineages linking us back to the 'mitochondrial Eve' the idea that all men descend from 10 individuals,[7] as if no other men lived at that time, is a product of the way the data are analysed. Although this approach bolsters our confidence, in that it comes up with roughly the same answer about the age of our species as the mtDNA analysis, molecular anthropologists readily acknowledge how speculative these hypotheses still are.

It has been argued that the models that exploit mtDNA and Y-chromosome data rest on shaky assumptions about DNA mutation rates and past human reproductive behaviour. These objections appear to have been countered by an additional aspect of dating changes in DNA, which depends on the analysis of what are called '*microsatellites*'. These consist of short repetitive sequences of two to five nucleotides. The most prevalent form contains only two nucleotides, cytosine and adenine (C and A), and therefore the segment is CACACACA..., which has been delightfully described by Luigi Luce Cavalli-Sforza as 'stuttering' (Calvalli-Sforza, 2000; p. 82). Errors often occur here when DNA is copied so that the number of repetitions increases or decreases in the new gene.

Microsatellites seem to be randomly scattered throughout the genome. Thousands have been identified. Their importance in terms of evolutionary studies is that they provide an improved method of genetic dating. In principle all forms of genetic dating depend on estimates of mutation rates and hence are independent of other estimates of dates based on other measurement techniques (see Appendix). Whereas the mutation rates for other parts of the genome are rare, that for microsatellites is high. This enables the mutation rate to be estimated with much greater confidence. Comparisons of

[7] Just as with mitochondrial Eve, the 'genetic Adam' was not the only man alive at the time, nor was he alive at the same time as Eve: it is that it is his form of Y chromosome that has survived in humans living today.

estimates of dates obtained by mtDNA, Y-chromosome and microsatellite methods have shown that broadly consistent results can be obtained from the different approaches. The greater precision obtainable from the microsatellite approach has, however, led to a narrowing down of estimates of when important human evolutionary developments took place.

3.8 WALKING OUT OF AFRICA

In spite of the continuing debate about the details, combining archaeology and genetics provides an increasingly consistent picture of the history of modern humans. Analysis of genetic markers points clearly to their evolving in Africa about 150 kya. Fossil forms similar to modern humans are first found in sites dated around 160 kya in Ethiopia (White *et al.*, 2003). The timing of the emergence of modern humans may have been a consequence of the challenges of the penultimate ice age (running from 220 to 130 kya: see Fig. 2.4). This glacial period was, if anything, more sustained than the most recent one. This probably led to extreme desiccation across much of Africa, especially during its final stages, and this could have led to a collapse in the population of archaic humans living in the continent at the time, the result of which was that only the most inventive or adaptable survived.

At this time Europe was inhabited by Neanderthals and the Far East by humans that were probably descendants of *Homo erectus*. The fossil record then suggests that around 100 kya modern humans began to spread farther afield. The traditional view is that modern humans colonized the Old World from the Levant. This appears to be consistent with the presence of undisputed *Homo sapiens* fossils dated to around 100 kya from Skhul and Qafzeh in what is now Israel. Here there may have been an overlap with Neanderthals, who appear to have occupied the region at times between 100 and 60 kya. There is, however, an unresolved debate about whether this initial diaspora was sustained or died out. The climatic evidence points to this first exodus failing (Oppenheimer, 2003; p. 54). In this context, the prolonged cooling from around 100 kya, which culminated in the cold period around

87 kya, may have made much of the Levant uninhabitable. This cooling may also have driven Neanderthals southwards from Eurasia, eventually reaching a deserted Middle East. The net result of this sequence is that there may have been no time when modern humans and Neanderthals simultaneously occupied the region.

If the presence of morphologically modern humans in the Levant around 100 kya was not part of the successful exodus out of Africa, then we have to look elsewhere for the evidence of the movement. The recent discovery of artefacts in Eritrea's Red Sea coast dated around 125 kya has shifted attention toward the possibility of a southern route out of Africa (Stringer, 2000; Walter *et al.*, 2000). The important question is when these people chose to cross the Red Sea. As we will see, the genetic evidence suggests that this departure did not occur until around 80 kya. So the next question is whether there is any obvious climatic trigger for such a movement. Here, the drop in sea level after 85 kya may provide the answer. The narrow Strait of Bab-el-Mandab at the mouth of the Red Sea acts as a 'pinch point' controlling the amount of water flowing between the Red Sea and the Gulf of Aden. As the sea level fell, not only would this narrow stretch of water have become easier to cross, but also the amount of water flowing into the enclosed sea would drop sharply. Analysis of sediment cores from the Red Sea show that as sea level fell the salinity of the water rose and the amount of plankton dropped (Rohling *et al.*, 1998). So, the rapid drop in sea level from the relatively high stand of 25 m below current levels that started around 84 kya (see Fig. 2.11) may have been the trigger. The decline continued until 76 kya, by which time the level had fallen by some 40 m. This would have had a catastrophic effect on marine life in the Red Sea. At some stage during this transformation the green hills of Yemen may have looked a more attractive prospect, and our ancestors crossed the narrow chain of islands to the other side.

After Arabia, humans could have easily moved along the coastline to India and Indonesia, and then made the more heroic leap to Australia. This shoreline expansion, hopping across narrow stretches of sea, would fit particularly well with claims that modern humans had arrived in Australia before 60 kya. There is, however, a continuing debate about the timing of their arrival. While there is a considerable amount of archaeological evidence of human occupation of sites in northern Australia dating around 60 to 55 kya, unequivocal dating of human remains has been more controversial. A dating of 62 kya for a skeleton found in a burial site at Lake Mungo, in southeastern Australia, has been overtaken by more recent analysis (Bowler *et al.*, 2003) that suggests a much later date of around 40 kya, although humans were present at Lake Mungo by 50 to 46 kya. Furthermore, there is considerable evidence of mass extinctions of large animals around 46 kya that has been linked to the arrival of modern humans (see Section 5.6).

The genetic and climatic data provide further clues to add to the debate on the timing of the arrival of modern humans in Australia. The genetic data are equivocal. Initial results suggested that there was a marked difference between the profiles for descendants of the earliest humans in Australia and those living in New Guinea. More recent analysis (Ingman & Gyllensten, 2003) shows that, while most of the lineages appear to be separate, there are two lines that show an admixture of groups from the two regions. These results can be interpreted as showing that the populations of the two regions could have emerged from a single colonisation at some time between 70 to 40 kya. Alternatively, the common threads in the genetic data may be the result of two separate colonisations followed by some movement between New Guinea and Australia. This later mingling of people could have taken place at any time until around 10 kya as the two regions were linked by a land bridge (the Sahul Shelf) until then.

If the principal colonisations were separate then the people who first travelled to Australia must have crossed the Timor Sea, rather than going via New Guinea. In these circumstances, known changes in the climate may provide insights into the timing of this movement. Given the distance involved, the most likely time for this trip to be completed was when the sea level was particularly low, as this would have exposed much more continental shelf to the north of Australia. The most opportune time would have been after the Heinrich 6 event when the level was around 70 m lower than current levels, say around 67 to 61 kya.

The combination of mtDNA and Y-chromosome studies has produced a similar picture of the move out of Africa, but with an important refinement to the chronology. Measurements of mtDNA diversity of 62 human population samples have provided one set of figures. Assuming an mtDNA divergence rate of 33% per million years, most populations show signals of a split at around 70 kya between Africa and Asia, and then around 55 kya between Asia and America, and around 40 kya between Europe and the Middle East.[8] In terms of the climatic template, it would appear that migration into Eurasia had to wait until well after the long stadial that lasted until around 59 kya. The reason for this delay is that not only did the influence of the cold climate prevailing across most of Eurasia extend to the Middle East, but also much of the region was desert at this time. So rather than working their way northwards along major rivers, such as the Euphrates, Tigris and Indus, our ancestors stayed close to the warmth of the Indian Ocean.

The Y-chromosome data suggest a later move out of Africa centred around 60 kya, reaching Central Asia by 40 kya and then into Europe by 35 kya, but given the uncertainties in all the measurements, these figures must be regarded as effectively being in agreement.[9] Leaving aside the finer details of the conclusions of the genetic mapping work, which will be considered in Chapter 4, there is one immediate feature of the differences between the archaeological and genetic studies. This is that the genetic measurements provide convincing evidence that the last wave of colonisation out of Africa was the only successful one.

[8] Accessible presentations of these studies can be found in reviews such as Maca-Meyer *et al.* (2001), and Excoffier and Schneider (1999).

[9] This combines data from a variety of Y-chromosome studies, the most notable of which are Wells *et al.* (2001) and Ke *et al.* (2001).

There is one additional piece of information that may point to the final exodus occurring in the earlier part of the interval discussed above. This is archaeological evidence from Malaysia (Oppenheimer, 2003; pp 166–170). Stone tools that have been attributed to modern humans have been found below the layer of ash associated with the supervolcano Toba (see Section 3.5). This means that modern humans had reached Malaysia before this eruption around 71 kya. So they must have started out from Africa several thousand years earlier. Furthermore, this evidence is consistent with the possibility that the descendants of these people were in place to make the crossing to Australia during the period of low sea level between 67 and 61 kya.

The other implication of these observations is that those humans who had been part of the original diaspora from Africa, and who had remained in coastal plains of India and Pakistan, may have been wiped out by the impact of Toba. Unlike the Malaysian peninsula, which was right on the edge of the dust cloud, the Indian subcontinent was covered by a massive amount of ash. If this was what happened, then humans would have moved in from the west subsequently to fill the gap left by this depopulation. This hiatus would explain the marked genetic divide between those now living in the East Asia, New Guinea and Australia, and those living in India and farther west.

In summary, both archaeological and genetic data confirm that the history of human migrations is complex and cannot be generally represented as a series of large-scale, successful colonisations. At the same time, it is virtually certain that we have ruled out completely some component of genetic input from archaic *Homo* species, including the Neanderthals. The evidence continues to build an increasingly consistent case for the 'out of Africa' model.[10] This means that there is no genetic lineage between these archaic species living in Eurasia

[10] The original work on identifying the extent of the differences between the mtDNA of Neanderthals and modern humans was done by Svante Paabo's group at Leipzig (Krings *et al.*, 1997; 2000), and a recent review of the archaeological implications of the ancient separation of Neanderthals and modern humans can be found in Klein (2003).

before 100 kya and modern humans. There is much more to discover about how modern humans colonised Africa, and the rest of the world, including dating of growth and decline of their population from place to place and at different points in time. Nevertheless, there is growing support for the hypothesis that the last exodus dominates our story, in part because those humans had evolved a higher level of skills and consciousness than their predecessors (see Section 4.2).

In spite of this assertion, there is a fly in the ointment. There is still a fundamental problem with the underlying approach to genetic mapping. This is the whole question of population growth.[11] There is an assumption that natural selection took place in either a relatively steady population or one that was growing slowly. During the period 100 to 20 kya the population of modern humans may, however, have fluctuated dramatically in the face of various environmental threats, notably climate change. Because natural selection and effects of population size can be hard to tell apart, especially if the population underwent significant reversals, or even was wiped out in many places, the effect of natural selection could be amplified. In these circumstances, a favourable mutation can sweep through the population under the influence of natural selection. Focusing on the carriers of this favourable mutation, the process looks just like population growth: the number of carriers is small at first, then increases, and then levels off. For practical purposes, the two processes have identical effects on genetic variation. There is still no clean way of distinguishing them except by comparing DNA from different genetic loci. Steady population growth should affect every locus in the same way, whereas selection should affect different loci in different ways. There is evidence of disparate results from different loci that suggests that human genetic variation is influenced strongly by natural selection. This means that the application of genetic mapping to population history requires careful handling.

[11] Among the most accessible papers on population growth and bottlenecks are Harpending *et al.*, 1998, Wall and Przeworski, 2000, Harpending and Rogers, 2000, and Rogers, 2001.

In spite of these uncertainties, there is a broad consensus on the distribution of haplogroups and timing of their arrival in different parts of the world, which are set out in Figs. 1.4 and 3.6. This presentation provides wide margins on the timing of major migrations, and the debates surrounding these uncertainties will be explored more closely in the coming pages.

3.9 THE TRANSITION TO THE UPPER PALAEOLITHIC

So far the analysis of human existence has been in generalities. It is now time to turn to the firm evidence of life in the ice age and the role of climate change in this precarious existence. The transition to the Upper Palaeolithic occurred at a time when the climate in Europe was in fluctuating phase (see Fig. 2.6 and Table 2.1), the principal features of which were:

- the relatively mild period following the ups and downs of DO oscillations 15 and 14 that extended from around 53 to 49 kya (the Glinde interstadial);
- the cold period associated with Heinrich event 5 around 48 kya;
- the relative warmth of the period from 47 kya until 40 kya, in spite of the punctuations of DO events 12, 11, 10 and 9;
- the cold period associated with Heinrich event 4 around 39 kya; and
- the relative warmth of the period from 38 kya until 32 kya, in spite of the punctuations of DO events 8, 7, 6 and 5.

In emphasising the ups and downs of this period, and the suddenness of many of the climate changes, it is important not to lose sight of the timescales involved. The spells of climate change associated with DO oscillations are measured in terms of multiples of 1.5 kyr. It is as well to remember that it is less than 1 kyr since the Norman Conquest of Britain that encompasses the Medieval Climate Optimum and the Little Ice Age: enough time for quite a lot to happen. What is more, these recent climatic periods have been cited as the Holocene variant of the 1.5-kyr cycle that could have driven the DO oscillations (Bond *et al.*, 2001).

At this point, introducing a little standard terminology assists in linking the archaeological record and the description of features of the Middle and Upper Palaeolithic with the results of genetic mapping. These terms reflect the original work and the names of the principal sites that were used to establish the archaeological chronology. The standard stone tools of the late Middle Palaeolithic in Europe are known as 'Mousterian' and are attributed to Neanderthals, who were the sole human occupants of Europe until around 45 to 40 kya (Klein, 2003; Mellars, 2004).

When it comes to the Upper Palaeolithic, four principal technological stages have been identified. The first is 'Aurignacian': this term is used to define the emergence of more sophisticated implements around 40 to 35 kya, which has now become synonymous with the arrival of modern humans in Europe. They probably came from the Levant and Turkey, crossing the Bosphorus, which was above sea level at the time. The earliest known Aurignacian sites are in the Balkans, around 43 kya. Their initial movement seems to have been along the Danube valley. At the same time similar people were moving round the shoreline of the Mediterranean into Greece and Italy. Three thousand years later at the most, the Aurignacian technology appears across the continent in Spain. Within a few thousand years the technology covers most of the rest of Europe. The analysis of this spread is complicated by the fact that there is an overlap with examples of a more sophisticated version of the Middle Palaeolithic technology, which is known as 'Châtelperronian'. This development appears to be a refinement of the Mousterian and is attributed to Neanderthals, who may have been copying the work of their Aurignacian neighbours.

The second stage is 'Gravettian', a term that has been associated with the next obvious step forward in stone tool technology, which dates from 29 to 22 kya. This group appears to have moved into Europe between 30 and 35 kya, either from the Trans-Caucasus region beyond the Black Sea, or possibly from farther afield to east of the Caspian Sea (Oppenheimer, 2003; pp. 144–5). During the LGM the Aurignacian culture became restricted to refuges around the Pyrenees and the Ukraine, while the Gravettian culture survived in the Balkans. From these refugia came the people who were to repopulate Europe at the end of the ice age, and who are the ancestors of most modern Europeans.

The third group is 'Solutrean': this culture existed in southwest France from around 22 to 17 kya, and is marked by incredibly finely worked leaf-shaped stone blades and points, which seem to have elevated stone tools to the status of works of art. Finally, as Europe emerged from the ice age, the technology between 17 and 11 kya is known as 'Magdalenian'. These definitions give the impression of there being an obvious sequence in these successive groups, but when combined with the geographical spread of occupation, there are some major gaps in the chronology. Moreover, it would be misleading to assume that successive waves took over. For instance, the Aurignacian appears to have remained the prevalent culture in western Europe, while the Gravettian culture prevailed in more eastern areas. These details raise interesting questions about whether various groups in different parts of Europe managed to survive the LGM, and if so, where.

The evidence of life at the beginning of the Upper Palaeolithic comes from a wide range of sites across Eurasia. It also encompasses a mixture of Mousterian, Châtelperronian and early Aurignacian technologies. The evidence is, however, spotty, both in terms of the chronology and the geography. So we need to be careful about drawing sweeping conclusions about the way changes occurred. Indeed, the archaeological analysis is complicated by the much more intriguing questions of the transitions taking place at the time: in other words, the arrival of modern humans and the disappearance of Neanderthals. Here, the interesting geographical aspect of the movement of the first modern humans into Europe is that they appear to have followed two principal routes. These are the corridors from the Balkans along the Danube Valley and from the Levant, Turkey and Greece around the coast into Italy, southern France and Spain. The genetic evidence is that this movement started around 50 kya. At roughly the same time there was a migration along the northern coast of Africa. The extraordinary feature of these migrations is that there is virtually no sign that they joined up at the Straits of Gibraltar: the genetic evidence shows a sharp discontinuity between the populations of North Africa and Iberia, with any limited signs of bi-directional flow being restricted to recorded history (Bosch *et al.*, 2001). Even then, the flow has been small.

The movement into eastern Europe from either western Asia or Anatolia was along the major river valleys. For this reason, some of the best-known sites have been found in riverside locales (e.g. Kostenki and Borshevo beside the River Don in Russia), or in caves (e.g. Willendorf, in Austria, and Geißenklösterle and Vogelherd, in southern Germany). Unfortunately, while the riverside sites have evidence of lengthy occupation, many of the caves contained only a limited amount of material, indicating specialised and intermittent occupation. Indeed this limited evidence may point to a migratory lifestyle that was a vital feature of surviving the rigours of the ice age. This process may not have been a matter of moving long distances, but more a question of having a pattern of seasonal movements within a region. This could have involved shifting between pitched camps or exploiting more permanent shelters in the most desirable places at given times of the year.

As we move through the Upper Palaeolithic towards the LGM, the chronological and geographical gaps in the record become, if anything, more difficult to interpret. The increasingly inhospitable conditions of much of Europe, north of the Alps, drove humans out, and the remaining sites of lengthy occupation look increasingly like random islands in time and space. So it is not possible to draw a continuous thread between lengthy occupation in one place and subsequent occupations elsewhere. The limited evidence from the warmer parts of the world is equally fragmentary. This means that in presenting a brief review of the archaeological record, it is not possible to provide a sense of continuity, still less of 'progress'. Instead, the presentation will try to provide a sense of the many ways in which humans rose to the challenge of surviving the ice age.

3.10 SETTLING ON THE PLAINS OF MORAVIA

In what was the treeless wilderness of the central European Plain just before the LGM, extensive archaeological evidence has been found of lengthy settlement between roughly 27 and 24 kya. This seemingly more settled lifestyle might have been the product not only of growing sophistication of human social structures, but also of the slightly less chaotic conditions prior to the LGM. So, the ability of modern humans to create this more settled existence was probably the key to their survival during the LGM, especially in refugia where conditions were that little bit more bearable.

The sites of this occupation are at Dolni Vestonice-Pavlov in southern Moravia. It is situated at the confluence of two valleys that appear to have been on a migration route for woolly mammoth. As such it was an ideal place to prey on these huge beasts, or simply to scavenge those that fell by the wayside or were killed by other carnivores. The benefits of the location presumably led to the long-term occupation that appears to have enabled a considerable social structure to grow up. An excavated building shows that it was first dug out from a slope and then the roof was supported with timber set into postholes. The low walls were made of packed clay and stones.

Dolni Vestonice is the site of the earliest known examples of ceramic artefacts. Many carved and moulded images of animals and women, strange engravings, personal ornaments and decorated graves have been found in the clay soil. In the main hut, where the people ate and slept, two items were found: a 'goddess' figurine made of fired clay and a small and exquisitely carved statuette made from mammoth ivory of a woman whose face droops on one side. The goddess figurine has holes on the top of its head that may have held grasses or herbs. The artist scratched two slits that stretched from the eyes to the chest, which may represent the life-giving tears of the mother goddess. A kiln was found above the encampment in a small hut. Scattered around the oven were many fragments of fired clay. Remains of clay animals, some stabbed as if hunted, and other pieces of blackened pottery still bear the fingerprints of the artist.

Another fascinating feature of this site is that an impression in the clay floor of a hut that was burnt down provides unmistakeable evidence of woven material. Evidence of 36 samples of textile impressions was found on both fired and unfired clay (Soffer, Adovasio & Hyland, 2000). These artefacts were clearly made of plant rather than animal fibres. The types of woven materials included single-ply, multiple-ply, and braided cordage, knotted netting, plaited and wicker-style basketry. This shows that some 15 kyr before the coming of agriculture modern humans had developed the ability to plait threads, and to knot and weave them. This enabled them not only to produce fabrics, but also to make nets that could be used to trap small animals, thereby extending the scope of their diet.

3.11 LIFE ON THE MAMMOTH STEPPES OF ASIA

One of the most remarkable features of the last ice age is the success of living on the plains of Russia. While northwestern Europe became uninhabitable during the LGM, in Russia occupation of a number of sites from the River Don to eastern Siberia appears to have continued unabated (Vasi'lev *et al.*, 2002). The occupation reflects not only the ability of modern humans to adapt rapidly to harsh environments, but also the fact that the ice sheets were not so extensive over the Siberian Arctic. Even during the LGM the mainland of northern European Russia was free of ice because of the limited extent of the Barents–Kara Ice lobes of the Fennoscandian ice sheet. One consequence of this appears to have been that the warming of the interstadial period between 32 and 29 kya (see Table 2.1; the period is termed the Briansk in Russia) may have been greater in Siberia. So the short relatively warm summer sustained a relatively abundant flora and fauna of the mammoth steppe (see Section 3.4).

Farther east, the earliest known occurrences of modern humans in Siberia are dated to around 45–41 kya. These are concentrated mostly in the Altai Mountains. By between 35 and 30 kya they had spread into the region beyond Lake Baikal (known as *Transbaikal*) and the Mongolian Gobi, and they were in northwestern China by 25 kya. During the same period they moved into the Yenisei, Angara and Upper Lena River basins. An archaeological site on the Aldan River, a tributary of the Lena in eastern Siberia, was occupied by a possible ancestor group to palaeo-Arctic people of North America. The people who lived in Dyuktai Cave were hunter-gatherers and fishers and used triangular stone points that have become known as the 'Dyuktai culture'. Occupation levels have been dated between 33 and 10 kya, although there is some doubt about the earlier dates. The recent dating of artefacts in the Yana River region to 30 kya confirms, however, that modern humans had reached northeast Siberia by this time (Pitulko *et al.*, 2004).

Farther east, there is evidence that modern humans had lived in Japan from about 30 kya, based on dating their flint tools. All four main Japanese islands were connected, and the southern island of Kyushu was connected to the Korean peninsula while the northern island of Hokkaido was linked to Siberia. These people appear to have survived the ice age and then around 12 kya developed a unique culture, which lasted for several thousand years. Their culture is known as 'Jomon', which means 'cord pattern', to describe the design of the pottery that these people produced – the earliest in human history. What is remarkable is that the Jomon were still a hunting, gathering and fishing society, living in small groups, when they developed this advanced technology. Furthermore, they also fashioned ceramic figurines.

There are other sites in Siberia of similar age. For instance two similar sites at Mal'ta and Buret, both in the Irkutz district of Siberia, have been found to date from around 28–25 kya. These sites are famous for the dwellings constructed of large animal bones, and the presence of a number of human figurines. The use of mammoth bones to build huts is a feature of the steppes. The best-known example is at a later site, dating from 15 kya, in the Ukraine at Mezhirich, where the remains of

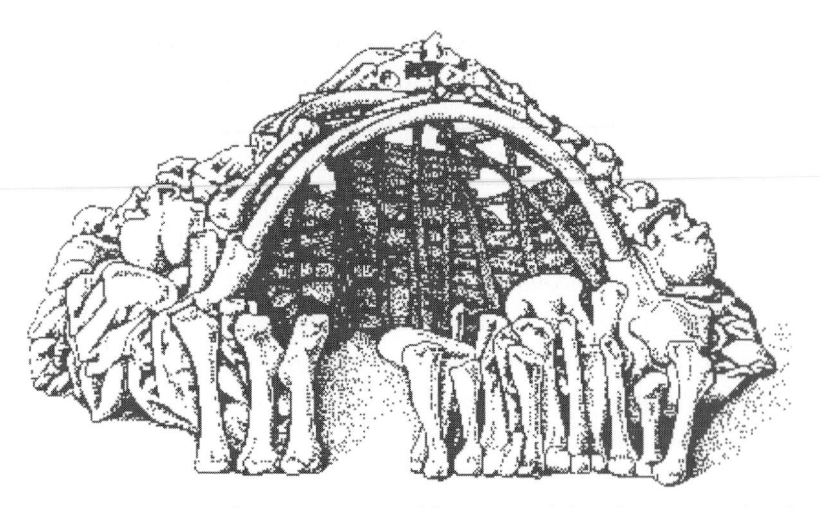

FIGURE 3.7. The reconstruction of the mammoth-bone house at Mezhirich. This reconstruction is in the Kiev Museum of Paleontology.

four huts consisted of complex arrangements of tons of mammoth bones. The layout of these bones has been defined as the 'earliest architecture'. One hut, some four to five metres across, had a careful herringbone pattern of mammoth lower jaws; another a palisade-like ring of long bones placed on end (see Fig. 3.7). It has been estimated that the total number of bones incorporated in the structure belonged to a minimum of 95 mammoths. This need not be a measure of the inhabitants' hunting prowess. Gnawing marks of carnivores on the bones suggest that many of them were scavenged. Nevertheless, dragging the enormous skulls across country was no mean feat, as even a small one weighed about 100 kg. Here, as at other Siberian sites, there is evidence that the inhabitants dug pits in the permafrost to store meat and bones: just like present-day point Barrow, they could then stay put, living off their reserves of meat, even when the migratory herds on which they depended were far away.

In terms of interpreting the creativity of these people, perhaps the most impressive site is at Sunghir, some 150 km northeast of present-day Moscow. Here are the spectacular burials of two children, aged about nine and thirteen, and a man, possibly as old as 60, dating from around 25 kya. All three bodies have been decorated with a phenomenal number of artefacts. These consisted of many thousands of mammoth ivory beads, several hundred fox teeth pendants and a range of ivory ornaments. The scale of the decorations shows that, at a time when the climate was in the depths of the LGM, the Gravettian culture in this exceedingly inhospitable region had sufficient spare time that it was capable of producing such finely worked and time-consuming items of personal decoration. It also suggests the society had an established system of social differentiation that attached considerable status to certain children as it is inconceivable they could devote such effort to producing decorations for every child in the community.

Farther south at Kostenki on the Don River about 400 km south of Moscow, where a series of more than 20 sites have been excavated, there is evidence of occupation by modern humans back to around 40 kya. Recent excavations have yielded bone and ivory needles with eyelets, dating from 30 kya. In addition, the research team uncovered neatly articulated bones of both arctic foxes and hares at the site. These discoveries suggest that residents of Kostenki had developed trapping techniques to obtain furs, which they sewed together to produce more effective clothing that would help keep them warmer in the winters. Animal remains found at Kostenki included horses, mammoth, bison, moose and reindeer. There also is evidence that they killed other small mammals and possibly birds using darts. In addition, analyses of bone chemistry from human remains provide evidence for high consumption of fish, representing another advance in human technology. The fishing requires control of the local environment with weirs or traps.

There is one other question about the occupation of Siberia during the LGM. It has been argued that much of the region was depopulated during the coldest period. It is probably inevitable that during the LGM there was a decrease in the area inhabited and that the population moved southwards. But, as cited above, there is considerable evidence that southern Siberia and the Russian Far East remained occupied throughout the LGM. The scale of this occupation is important as it affects the timing and continuity of the peopling of North America (see Section 5.12).

3.12 SHELTER FROM THE STORM

One of the fascinating puzzles about human occupation in Europe during the LGM is the location of preferred occupation. Often termed *refugia*, these regions were a matter of life and death. Finding a niche in the harsh environment may well have been a question of the survival of the luckiest. So far, the sites discussed have been in some pretty desolate places and it is not clear why they should have been chosen. The most obvious explanation is that they were well placed for exploiting major migration routes of prey (mammoth, reindeer etc.). This is not the whole story, because year-round survival depended on other predictable sources of food. So the related question of where flora and fauna also maintained a toehold is an equally interesting aspect of surviving the ice age. Put at its simplest level, humans were probably most able to survive where a range of plants and animals could also manage to adapt to changing conditions.

The archaeological record shows that humans survived the LGM in northerly latitudes in a number of favourable regions, notably in northern Iberia and southwest France, Italy, the Balkans and Greece. A good indication of what constituted a refugium for flora and fauna has emerged from studies carried out in the Pindus Mountains in northwest Greece (Tzedakis *et al.*, 2002). A high-resolution pollen record from the Ioannina Basin, at an altitude of 470 m, shows that the amplitude of millennial-scale oscillations in tree abundance during the last ice age was subdued. Although the major climatic fluctuations associated with OIS 5, 4, 3 and 2 show up clearly in the populations of different species, temperate tree populations survived throughout the glacial period. The evidence from this refugium is that it was an area of relative ecological stability, reflecting the influence of continued moisture availability and varied topography.

While refugia enabled flora and fauna to survive in place, for those species that could not migrate the climatic chaos of the ice age meant death. In the case of plants, migration took place by the normal process of seed dispersal, provided that the climatic shifts were not so rapid and substantial that none reached a climatically tolerable sanctuary. It was, however, in the nature of the ice-age climate that changes were too sudden for this response to work in many instances. Similarly, animals would have adapted their migration patterns, with the weakest going to the wall. In the case of humans the same icy logic would have applied. In the search for shelter from the icy blast, the most obvious solution, as some American anthropologists say, was for our ancestors 'to go to the beach'. This strategy had the obvious benefit that close to the ocean the climate was more moderate. The second advantage was that the seashore is an abundant source of food, notably in the form of shellfish. So, just as with the encampments that relied on migrating big game, people living in caves or other shelters by the sea had a more reliable source of food. The problem for us is that the evidence of their presence has been swept away by the rising sea many millennia ago.

3.13 THE FIRST FISHERMEN OF GALILEE

So far all the emphasis of living through the ice age has been an unremitting tale of grappling with almost insurmountable odds. This may not be too much of an exaggeration when compared with modern life, but it needs to be emphasised that in certain circumstances our forebears may have been able to live off remarkably abundant food supplies. The more temperate climate at lower latitudes, even allowing for the influence of the climatic ups and downs in the North Atlantic, would have been a benefit. What would have mattered most were the fluctuations in rainfall: when it was adequate life was pretty good, when it was parched there was no record and nothing left for us to see now. So, during the LGM, the archaeological record is the evidence of life at a time of temperate climate and relatively abundant rainfall.

Evidence of the ice-age lifestyle in the Levant was discovered at Ohalo in Israel in 1989 (Nadel & Werker, 1999), following a drastic drop in the water level of the Sea of Galilee. Excavations in 1989, and then again in 1999, when the water level fell sufficiently low, revealed the remains of a camp, including six huts, a grave, a stone installation, several kinds of fireplaces and what seems to have been an area for garbage disposal. The charred remains of the huts contained a wealth of material on the floors, including flints, animal bones and burnt fruit or seeds. The hearths were placed outside the huts, and provided further valuable information.

Carbon dating of these finds gives an average age of around 23 kya. All features at the site appear to belong to this one period, which falls within the LGM. This was a time of plentiful rainfall, soon after the extreme drought of Heinrich event 2 (see Section 3.5). The diet of the people occupying the site was extremely varied. Remains of tens of thousands of seeds and fruits of about a hundred species have been identified so far. These include many edible plant species, such as wild barley, wild wheat and acorns. Thousands of fish bones show that fish was central to the local economy. Furthermore, thousands of gazelle bones and numerous bones of fallow deer, fox, hare and other species indicate the other sources of food. Consideration of the ripening months of the recovered seeds and fruits, together with the analysis of gazelle teeth and the bones of birds, shows that the camp was used on a year-round basis.

Perhaps the most important aspect of the work on this site has been the recent analysis of starch grain residues found on a grinding stone (Piperno *et al.*, 2004). This provides evidence of processing of wild cereal grains. Associated evidence for an oven-like hearth on the site suggests that dough was made by baking grain flour. This processing and cooking of wild cereals some 12 kyr before the advent of agriculture in the region (see Section 5.11) raises intriguing questions about the existence of what is sometimes referred to as proto-agriculture, managed to exploit grain seeds at a time when the climate was too variable to support more organised agriculture.

A grave found to the west of the huts provides additional important insights into the nature of this community. There, a complete skeleton was found in a shallow pit. The burial was of an adult male, 1.73 m tall and 35 to 40 years old at the time of death. According to the arm and chest bones, he was disabled during his last years. This is interpreted as reflecting social commitment to such members of the local group. All of this raises interesting questions about the degree of social organisation among people living in coastal sites during the LGM that now lie below the waves.

3.14 WADI KUBBANIYA AND THE KOM OMBO PLAIN

In Egypt, evidence of human occupation around the end of the LGM has been found at a number of sites. At the time, the general aridity of the climate greatly reduced the flow of the river Nile. Indeed for much of the time, the White Nile flowing from equatorial Africa disappeared into the dunes of the Sahara in the region that now constitutes the great El Sudd swamplands in southern Sudan. The reduced flow of the Blue Nile from the Highlands of Ethiopia, which were grasslands, carried much more sediment than now. So, along much of its length, the Nile Valley built up silt to some 30 m above current levels.

The best-known site is at Wadi Kubbaniya, a dried-up streambed cutting through the Western Desert to the floodplain northwest of Aswan in Upper Egypt (Wendorf *et al.*, 1979; Wendorf *et al.*, 1984). A cluster of camps was located on the tops of dunes and the floor of the wadi where it enters the valley. Although no signs of houses were found, diverse and sophisticated stone implements for hunting, fishing, and collecting and processing plants were discovered around hearths. Most tools were bladelets made from chert. The bones of wild cattle, hartebeest, many types of fish and birds, and even the occasional hippopotamus have been identified in the occupation layers. Charred remains of plants that the inhabitants consumed, especially tubers, have also been found.

It appears from the zoological and botanical remains at the various sites in this wadi that the two environmental zones were exploited at different times. We know that the dune sites were occupied when the Nile River flooded the wadi because large numbers of fish and migratory bird bones were found at this location. When the water receded, people then moved down onto the silt left behind on the wadi floor and the floodplain, probably following large animals that looked for water there in the dry season. Palaeolithic peoples lived at Wadi Kubbaniya for about two thousand years, exploiting the different environments as the seasons changed. Other ancient camps have been discovered along the Nile from Sudan to the Mediterranean, yielding similar tools and food remains.

Further evidence of human habitation of the Nile Valley at the end of the LGM has come from the Kom Ombo Plain (Smith, 1976), which is a rich alluvial plain 50 km north of Aswan. Between 17 and 12 kya this area offered an attractive habitat for humans. Rainfall, having increased at the end of the LGM, was more abundant than now. So not only were the Nile floods more substantial, but also the rainfall in the Red Sea Hills to the east of the river was sufficient to feed the now dried-up tributaries that ran into the Nile across the Kom Ombo Plain. The range of foods was substantial. Animal bones included a now extinct large wild ox, the bubal hartebeest, several species of gazelle and hippopotamus, which appeared to be the principal game eaten. In addition, there were hares, hyenas, a form of dog, bandicoot rats and possibly 'Barbary' sheep. The streams and pools provided Nile catfish, Nile perch, the African barbel, and local species of oyster and soft-shelled turtle. The bones of some 22 forms of birds were also identified. In addition, the food supply would have included roots and bulbs throughout the year, together with seasonal supplies of berries, nuts, and perhaps melons and cucumbers plus edible gums from various trees.

These sites demonstrate that the early inhabitants of the Nile valley and its nearby deserts had learned how to exploit local environments. In the case of the Kom Ombo Plain the area appears to have been abandoned when the climate became much more arid, possibly in association with the onset of the Younger Dryas. Within the Nile Valley changes associated with the climatic fluctuations following the LGM and the early Holocene were complicated. The greatly increased rainfall both in the headwater regions and also along the river that occurred around 14.5 kya (see Fig. 2.11) opened up the White Nile and had the counterintuitive impact of scouring a deep channel along much of the valley in Egypt. This narrow flood-prone region was less attractive to humans than the surrounding hills and wadis. Following the drier interlude associated with the Younger Dryas, the torrents returned, making the valley less hospitable then the surrounding savannah . In some places, a hunting and fishing lifestyle continued along the Nile Valley in Lower Egypt and Nubia, but it would not be until the drier conditions of the mid-Holocene that people returned in large numbers to the valley (see Section 5.15).

3.15 THREE-DOG NIGHTS

The Chukchi, who live in the far east of Siberia, are renowned for having bred the Siberian husky. This dog is legendary for its strength, stamina and sweetness of temperament. For as long the Chukchi can recall their life has revolved around huskies. The women reared the pups and chose which pups to keep, discarding all but the best bitches and neutering all but the best males. The men trained the dogs, principally the neutered males, to pull sledges. Huskies also acted as companions for the children and families. In the bitterly cold Siberian winters the dogs slept inside to keep people warm. Chukchi folklore defined temperatures at night in terms of the number of dogs necessary to keep a person warm: a cold night was 'two-dog night' and an intensely cold night a 'three-dog night'. This raises the interesting question as to when the extraordinarily close relationship between dogs and modern humans was first established.

Geneticists have used mtDNA analysis to investigate the timing of domestication of various animals. One of the most controversial examples of this type of work was done by Carles Vila and co-workers at the University of California at Los Angeles, the Royal Institute of Technology, Stockholm, Sweden, Texas A&M University and Brigham Young University, Provo, Utah, on the multiple and ancient origins of the domestic dog (Vila *et al.*, 1997). The study analysed mtDNA control-region sequences from 162 wolves at 27 localities worldwide and from 140 domestic dogs representing 67 breeds. Sequences from both dogs and wolves showed considerable diversity and supported the hypothesis that wolves were the ancestors of dogs. Most dog sequences belong to a divergent common line from a single ancestral species (*monophyletic clade*) sharing no sequences with wolves. The sequence divergence within this clade suggests the dogs originated more than 100 kya. There is, however, evidence that there has been more recent admixture between wolves and dogs, and that this may have been an important source of artificial breeding of dog species.

The implications of this work are fascinating. If there was some form of symbiotic relationship between humans and wolves that led to the appearance of dogs as long ago as the early stages of the last ice age it could have profound consequences for our interpretation of life during this glacial period. This proposal does not necessarily fly in the face of the more standard view that dogs were not truly domesticated until the end of the ice age.

In the context of life in the ice age, domestication is probably too strong a word to use. Nevertheless, there is plenty of evidence of the bones of wolves being found in association with those of hominids as long ago as 400 kya. Equally certain, until recently, was that remains of morphologically distinct dogs did not appear in the archaeological record until around 14 kya, at the earliest, in the Middle East. Recent fossil evidence from the Russian steppes has pushed this date back in time. An archaeological dig in the Bryansk region of Russia found the remains of two large dogs with intact skulls (Sablin & Khlopachev, 2002). Dated to between 16 and 20 kya, these dogs had lived in the severe conditions at the end of the LGM. Their bones were found next to those of the mammoth, polar fox and reindeer. Their physique resembled that of enormous huskies, but they were shorter and wider in the muzzle than the contemporary huskies. They would have been well suited for hunting and guard duty. In addition, the presence of holes cut in the skulls to remove their brains is evidence that they were also a source of food.

This discovery suggests that wolves had already undergone considerable selection, if not domestication, by palaeolithic hunters by the end of the ice age. The traditional view – that the bones of wolves in archaeological sites dating from earlier times were simply evidence of our voracious ancestors eating them – may not then be the whole story. Instead, they may, in some cases, have been the remains of the predecessors of early dogs that were part of the social group, although they may have ended up on the menu when times were hard.

In the harsh world of the ice age the mutual benefits accruing to both humans and wolves of hunting and scavenging together may have become apparent at an early stage. Inevitably this would have started as a competitive process, but at some stage it could have evolved into something more complicated. Ice-age hunters would have sought to exploit the wolves' ability to find prey and their teamwork in attacking and killing large animals. At some stage young wolves might have been rescued and raised as pets to help in understanding these skills. Throughout this emerging symbiosis there would have been little reason to seek to change the form of the semi-domesticated wolf or proto-dog. So, although this species would have evolved while being part of the early human society its physique would not have changed appreciably as it was already superbly well equipped to do what ice-age hunters wanted it to do. Only with the coming of agriculture and the formation of large settlements would the possibility of selective breeding regimes on dogs have led to marked changes from wild wolves.

These issues have become the subject of debate recently, as the latest studies on dogs suggest that the separation from wolves may have come later (Savolainen *et al.*, 2002). The figures in the latest work indicate a range from 15 to 40 kya, with a preference for the more recent date. Nonetheless, work published at the same time by Jennifer Leonard, Carles Vila and colleagues supports the hypothesis that ancient American and Eurasian domestic dogs share a common origin from Old World grey wolves (Leonard *et al.*, 2002). This implies when humans colonized America, at the latest 14 kya (see Section 5.12), they brought multiple lineages of domesticated dogs with them. The large diversity of mtDNA lineages in the dogs that colonized the New World implies that the ancestral population of dogs in Eurasia was large and well mixed at that time. Consequently, dogs, in association with humans or through trade, spread across Europe, Asia and the New World soon after they were domesticated.

The alternative is that domestication was a more ancient event, as suggested by previous genetic results. Furthermore, if the migration of people and dogs into North America took place at an earlier date, say 25 kya (see Section 5.12) then domestication would need to have taken place before this date. Even if a date of 15 kya for first domestication were accepted, the dog, as an element of culture, would have had to be transmitted across Paleolithic societies on three continents in a few thousand years or less. Such an extensive and rapid intercultural exchange during the Palaeolithic seems unlikely, and so there remains a strong case for humans and dogs having some form of symbiotic relationship during the later stages of the ice age.

In terms of life in the ice age, an early, close relationship with dogs could have played a vital role in the survival of modern humans in Eurasia. The combination of humans and dogs makes a far more powerful hunting force than either species on its own. Is it too whimsical to go further and ask whether this relationship enabled the first humans in Beringia to survive the more frequent three-dog nights of the coldest episodes of the ice age?

3.16 OF LICE AND MEN

If the lengthy connection between dogs and humans seems a little farfetched, recent work on body lice is even more extraordinary. A study by molecular anthropologist Mark Stoneking, who worked with Rebecca Cann and Allan Wilson on the original 'mitochondrial Eve' study (see Section 3.7), and colleagues at the Max Planck Institute for Evolutionary Anthropology in Leipzig, Germany, into the mtDNA of the body louse, has identified when this species evolved (Kittler,

Kayser & Stoneking, 2003). Only humans carry this particular species of louse, which lays its eggs in clothing. Experts agree that body lice are a subspecies of head lice and that body lice probably evolved when people started to wear clothing. The results of the study suggest that the evolution of body louse occurred around 70 kya. The inference from this result is that around this time humans first started wearing clothes.

Yet again, we have another intriguing example of apparently important changes in the human condition taking place around 70 kya. It is argued that only when humans moved out of Africa and experienced colder climates did they start to wear clothes. Although it would be unwise to attach too much importance to this specific date, it could also be argued that the dramatic cooling that took place because of the supervolcano Toba might have stimulated the wearing of clothes wherever people were living. Whatever the real explanation, the climatic upheavals that occurred at this time could well have led to radical changes in how modern humans lived.

Equally remarkable is recent work (Reed *et al.*, 2004) that shows that there are two distinct genetic lineages of head lice. Despite looking virtually identical, these two forms of louse appear to have separated around 1.2 million years ago. This coincides roughly with the time when the archaic humans *Homo erectus* first ventured out of Africa. These humans lived in East Asia until about 50 kya, and the only way modern humans could have picked up their form of louse is by some social contact. This raises fascinating questions about the nature and timing of the interaction.

The geographical distribution of the two forms of louse is also interesting. One is found on people all over the world. The other is almost exclusive to the Americas. This suggests that the modern humans that crossed Beringia into the Americas carried the form of the louse that had survived for so long on our human cousins *Homo erectus*. What is more, the genetic patterns of the worldwide form of the louse show evidence of a bottleneck in their population around 100 kya, apparently in harmony with the tribulations of their human hosts (see Section 4.1).

All of this confirms the extraordinary range of sources being used to tell us more and more about our past. This progress is summed up by the delightful connection made by *Current Biology*, when it published Stoneking and colleagues' paper: the cover carried the banner 'Of lice and men'. What better point to bring this discussion of life in the ice age to a close?

4 The evolutionary implications of living with the ice age

Man . . .
Who trusted God was love indeed
And love Creation's final law –
Tho' Nature, red in tooth and claw
With ravine, shriek'd against his creed –

Alfred Lord Tennyson (1809–1892), *In Memoriam, 1850*

The expression 'red in tooth and claw' is metaphor often used to depict conditions during the Stone Age. Moreover, Tennyson's anguished cry against the brutality of life is often seen as a riposte to Darwinism and the whole conception of evolution. In truth, although Tennyson was deeply hostile to Darwin's evolutionary theories, he wrote his famous poem as an intensely personal expression of his sense of loss at the death of his dear friend A. H. Hallam, and this was first published nine years before Darwin published his greatest work, *On the Origin of Species*. Nevertheless, the powerful image of the awful inevitability of evolutionary processes provides a good point to embark upon the task of exploring how climate change influenced the evolution of modern humans.

The emergence of modern humans from Africa is a good starting point for considering the evolutionary implications of the ice age for our species. Clearly this choice is arbitrary. But, as has been made clear in the preceding chapter, our knowledge of both the timing of this movement and its details in the early stages is at best fragmentary. This uncertainty is not important; what is essential to the discussion is the fact that whenever we choose to start considering the development of modern humans, the vast proportion of this time was spent grappling with the fearful challenges of the ice age. So, as with the climatic discussion, we will begin at 100 kya, which means that some 90% of our consideration of the history of modern humans is during the last glacial.

The most basic issue we now have to face, at any given time, is how many modern humans were living through the ice age and where they were living. As we will see these numbers exert a profound influence on the interpretation of the evidence we now have of human evolution through the ice age. In particular, the whole process of genetic mapping (see Section 3.7) has to come to terms with this question. It is implicit in the discussion here, however, that fluctuation in populations is even more crucial to the analysis. Inevitably, we have to work on the basis that the most extreme examples of climate change are likely to have precipitated demographic crises, especially in those regions already experiencing challenging conditions. So we must now look at the underlying metaphor for these crises, namely 'bottlenecks', which describe the nature of the constriction that the human race survived.

4.1 BOTTLENECKS

The whole discussion of how the population varied during the ice age is an area where there are more theories than evidence. This is hardly surprising as there are no direct measures of population numbers. Moreover, as noted in Section 3.8, fluctuations in population levels expose the differences between the various genetic mapping techniques (see for instance Excoffier and Schneider, 1999; O'Connell, 1999; and Harpending and Rogers, 2000). Clearly, at the local level, population fluctuations have exerted a major influence on the archaeological record. Rises in population are inferred by quantifying measures of increasing social and economic activity, which are reflected in the growing size and number of settlements. On the other hand, depopulation is measured in terms of declining activity and the abandonment of settlements, whether as a consequence of warfare or a breakdown in social order. Integrated over wider areas, these measures can be built up into some form of estimate of population. These changes are implicitly connected to fluctuations in the climate, but the nature of these links is at best speculative.

Although these changes form part of the analysis of the impact of climate change in prehistoric times, there is a more important issue to consider. This is the question of longer-term changes in population that underpin both the archaeological and, perhaps more importantly, the genetic history of our species. These fluctuations can operate in either direction. A sustained improvement in living conditions can lead to a growth in population over, say 10 to 20 generations, which leads to migration (e.g. the arrival of modern humans in Europe at the beginning of the Upper Palaeolithic). Evidence of these movements, and by implication population pressures, is found in the form of remains and artefacts in the archaeological record.

Conversely, a sudden deterioration in the climate or the emergence of a virulent new form of disease can lead to a sudden decline in numbers. In prehistory the evidence of such catastrophes is hard to see, as it leaves no bits and pieces. A marked decline in artefacts or other evidence of occupation may provide subtle clues that communities were being pushed closer to the edge of oblivion. But there is rarely a well-established order in how these things take place. So, when the evidence of human activity disappears from the archaeological record, the nature and timing of the cessation of this 'progress' is rarely evident. It is hardly surprising that we have no direct measures of population numbers and can only extrapolate from the environmental record to make the most rudimentary estimates.

The speculative nature of population estimates has direct consequences for genetic mapping. This discipline has to make some assumptions not only about the rate at which genetic mutations arose but also what forms of natural selection occurred that may have accelerated the pace at which some mutations became established. Fluctuations in human populations can have a huge impact on the rate at which natural selection operates. This leads to a phenomenon known as *genetic drift*, which refers to changes in the genetic composition of a population due to a dramatic reduction in breeding numbers. When numbers fall to very low levels, simple chance may allow some genetic variants resulting from mutations to spread throughout the population, whereas others die out altogether. In this extreme form, this is known as *random genetic drift* or *founders' effect*. In these circumstances, the survival of the fittest, or, as some molecular anthropologists have observed, 'the survival of the luckiest', has profound effects on the genetic record (Cavalli-Sforza, 2000; p. 42).

What geneticists expected to see was a pattern of markers that was consistent with a relatively constant population size. Instead, what they saw was a pattern much more consistent with a dramatic reduction in population size at some point in our past. Given the length of time humans have existed, there should be a wide range of genetic variation, yet DNA from people throughout the world is surprisingly similar. One answer to this conundrum is a bottleneck. This has led anthropologists and geneticists at the University of Utah to develop the 'Weak Garden of Eden' model for the origin and dispersal of modern humans (Harpending *et al.* 1993). This model proposes that, prior to the bottleneck, there was a much larger population of modern humans, who may or may not have been restricted to Africa. Then, as a result of the decline in population, the modern humans, who had only survived in Africa, split into a small number of geographically separate groups. As the population grew again within these separate groups, each starting from a tiny base, members of one of the groups set off to people the rest of the world. Although this expansion may not have marked the end of bottlenecks at the local level, it is of particular importance, because it marked the point when modern humans ultimately started to colonise the rest of the Earth.

What caused bottlenecks? The obvious candidate here is extreme climatic change. It is easy to postulate that sudden changes, associated with DO oscillations and Heinrich events, led to demographic crises. The fundamental question is, however, whether the genetic evidence points to any particular event or events. Sadly, no. It is not possible to date the genetic changes with adequate precision to point categorically to a given event. What can be said is that the bottleneck appears to have occurred around 70 to 80 kya, give or take 5 to 10 kyr. While this is a broad-brush figure, it does point towards one particular culprit. As discussed in Section 3.2, arguably the most dramatic of the cooling events in the last ice age occurred between 75 and 70 kya. In particular, the event that has been linked to the massive eruption of Mount Toba in Indonesia (Rampino & Self, 1992; Leuschner & Sirocko, 2000) at 71 kya is a likely candidate. This supervolcano could have produced what is sometimes referred to as a 'volcanic winter' that lasted several years and could conceivably have significantly altered global climate for the next 1000 years.

The dramatic cooling resulting from this eruption could have brought famine and death to many modern humans, leading to an abrupt decrease in our ancestors' populations: in short, a bottleneck. Anthropologist Stanley Ambrose of the University of Illinois has proposed that their numbers were reduced to levels low enough for evolutionary changes, which occur much faster in small populations, to produce rapid population differentiation, or genetic divergence of the surviving populations (Ambrose, 1998). The genetic evidence can be interpreted as indicating that the global human population crashed to no more than 15 000 to 40 000 people, and possibly lower.

This timing of the particular bottleneck has been linked with the exodus from Africa. Furthermore, if the surviving population were in Africa, this would be consistent with the mtDNA and Y-chromosome data (see Section 3.8) that points to the departure of the conclusive wave of modern humans from Africa between about 80 and 60 kya. In interpreting this range, however, we have to deal with the evidence that modern humans had reached Malaysia before the eruption of Toba (Oppenheimer, 2003; p. 168). The impact of any bottleneck at this time thus depends on whether all modern humans still resided in Africa, or had already started the long process of colonising the world. On balance, it appears that exodus had already started and that the main consequence of Toba was to separate those who had already reached Indonesia from those who were still in Arabia (see Section 3.8). Thereafter, the pace of migration was probably governed by the climatic conditions at the time. The fact that the mid-latitudes of the northern hemisphere were gripped by extreme glacial conditions, and the Middle East was extremely arid, for much of the period from 70 to 60 kya would have prevented any rapid expansion northwards. Only with the sustained onset of the relatively benign climatic conditions that started around 58 kya (see Table 2.1) would the gateway to Asia and Europe have become an attractive option.

Other factors would also have slowed the pace of colonisation. Initially, although the migrating branch of modern humans could well have undergone appreciable growth in what was a genetically isolated, small population, there simply would not have been enough of them to fill the void. Combined with the climatic barriers this slow growth helps explain the relatively late arrival, around 50 kya, of the modern humans in the Middle East, who then first moved into southeast Europe and southern Siberia by around 45 kya. Moreover, if climate change between 75 and 60 kya had a significant impact on human population levels and geographical distribution, then it follows that subsequent climatic ups and downs also affected population numbers. The fact that these culls are not as evident as the initial one suggests that their consequences were less marked.

In Eurasia it is reasonable to assume that between the interstadials and stadials of the Upper Palaeolithic the population rose and fell appreciably. Genetic studies provide little evidence of major fluctuations but do indicate a bottleneck during the LGM (Richards *et al.*, 2000). Thereafter, numbers grew steadily into the Holocene. In contrast, Amerindian and present-day hunter-gatherer populations show no evidence of this early growth. One explanation could be that there was some fundamental difference in the challenges facing hunter-gatherer populations in the ice age as opposed to the Holocene. Examination of present-day hunter-gatherer populations can be presented as being a series of recurring founder events and population crashes. So the issue is whether present-day hunter-gatherers provide a useful model for the past or whether the demographics of the ice age and the Holocene were drastically different (Excoffier & Schneider, 1999; O'Connell, 1999). Stable hunter-gatherer demographics during the ice age imply relatively large effective population sizes. This stability could have been achieved through high migration rates among subpopulations. A reduction of hunter-gatherer effective population sizes in the Holocene does not necessarily imply a drastic reduction of their absolute census size; it could be a consequence of fragmentation of the environment.

4.2 THE UPPER PALAEOLITHIC REVOLUTION

We have already touched on certain aspects of what is often termed the 'Upper Palaeolithic Revolution' (Mellars, 1994, 2004; Bar-Yosef, 2002). Now we must look closely at what the expression means. To call it a 'revolution' is probably an overstatement. What is evident is that the transition from the Middle Palaeolithic to the Upper Palaeolithic around 40 to 35 kya represents one of the major developments in the prehistory of humankind. The basic features of this transition included more versatile stone implements, the use of antler, bone and ivory for tools, figurative art, music and personal decoration. The best-known form of this creative activity is the cave paintings. This was, however, preceded by other forms of 'mobiliary' art, including various forms of representations of animals, and in one case a human figure with a lion's head from the Vogelherd Cave in southern Germany. The other famous form of mobiliary art is the 'Venus figurines' that were a widespread feature of Gravettian culture for the narrow time range around 25 to 23 kya.

Equally interesting is the emergence of music. The first unequivocal evidence of human musical activity (d'Errico *et al.*, 2003), in the form of a pipe made from a swan's bone dating from around 39 kya, was found at the Geißenklösterle Cave in Germany. More than 20 similar carefully crafted pipes, from the Aurignacian to Magdalenian periods, have been found at the Isturitz Cave in southwest France. Some of these pipes were apparently designed to be played two-handed, and have chamfered holes to maximise acoustic efficiency. They appear also to be polished by frequent use. So throughout the worst of the ice age humans may have used music to while away the cold dark nights.

The central issue to address concerning these developments is whether the transition identified in Europe marks the start of an entirely new form of human behaviour. Until recently it had been argued that this transition was an archaeological phenomenon found only in Eurasia. The apparent lack of equivalent evidence in other regions suggested that a fundamental change had occurred in human intellectual development around 40 kya in Europe. The recent discovery in the Blombos Cave in South Africa of a block of decorated ochre and then sets of shell beads, dated to around 77 kya, opened up the debate (Henshilwood *et al.*, 2002; Henshilwood *et al.*, 2004). This supports other evidence of more versatile stone implements and bone tools found in Africa from the same period. Now, the Upper Palaeolithic Revolution is being seen as simply the most visible example of the evolving process of modern human behaviour that had been developing over a much longer timescale.

This raises two further questions. First, what was happening to the human cognitive process during the 40 kyr or so between the creations in the Blombos Cave and the flourishing of human creativity in Europe around 35 kya, and was there a climatic component to this hiatus? The second is what stimulated the sudden efflorescence of creative activity at the beginning of the Upper Palaeolithic.

The question of whether the sudden transition seen in Europe was built on earlier developments in Africa has been addressed at length by Sally McBrearty, of the University of Connecticut, and Alison Brooks, of George Washington University (McBrearty & Brooks, 2000). They argue that the whole issue of the Upper Palaeolithic Revolution stems from a profound Eurocentric bias and a failure to appreciate the depth and breadth of the African archaeological record. In fact, many of the components of this revolution are found in the African Middle Palaeolithic tens of thousands of years before they appeared in Europe. These features include blade and microlithic technology, bone tools, increased geographic range, specialised hunting, exploitation of aquatic resources, long-distance exchange networks, systematic processing and use of pigment, and art and decoration. These items do not occur suddenly together as predicted by the revolutionary model, but at sites that are widely separated in space and time. This suggests a gradual assembling of the package of modern human behaviours in Africa, and its later export to other regions of the Old World.

The extraordinary range of rock art in Australia adds great weight to the idea that artistic creativity was part and parcel of the intellectual capacity of modern humans that migrated out of Africa around 70 kya. The fact that these people almost certainly arrived in Australia before 60 kya, and were, in any case, completely isolated from any evolutionary events that may have occurred in Europe around 40 kya, makes this argument compelling.

The consequence of this analysis is that the question of the sudden emergence of creative activity that appears to constitute the Upper Palaeolithic Revolution falls to the ground. The obvious explanation is that the gap between African developments and the subsequent better-known European events is a matter of the limitations of the archaeological record. This does not altogether cover the question of why there was the sudden flowering of creativity at the beginning of the Upper Palaeolithic in Europe. It may be that earlier creative efforts have either been lost in or have yet to emerge from the mists of time. Recent finds of decorative pierced shells in caves in Turkey and Lebanon, dating from 43 kya or even earlier, may be examples of a process of extending the evidence back in time (Kuhn *et al.*, 2001). The creative flowering may also be a result of the climatic conditions at the time that governed the movement of modern humans into Europe (Conard, 2003). Following Heinrich event 4 around 39 kya, the warming associated with DO event 8 (see Table 2.1), the region became more hospitable. As our ancestors moved into a largely depopulated region their presence in the archaeological record appeared revolutionary.

Other issues that swirl around the Upper Palaeolithic Revolution include the biological arguments about the role of modern humans in the demise of the Neanderthals. In addition, there are the cultural, technological, and environmental arguments about what constituted the most important changes. All of these issues are bedevilled by arguments about terminological ambiguities, and uncertainties about where and when changes took place. Overarching all of this is the hotly debated and loaded issue of whether these characteristics can be regarded as modern behaviour (Bar-Yosef, 2002). All of this suggests that improved knowledge of the parallel changes in climate will play little part in resolving these issues.

4.3 EUROPEANS' PALAEOLITHIC LINEAGE

Beneath the issues of bottlenecks and the evolution of human behaviour lies the more basic question of where and when various groups of humans separated. This has been the subject of particular study in Europe. Assuming that the modern humans who entered Europe came from Anatolia around 45 kya, there are two principal questions about their subsequent movements. First, where did they survive the LGM? Second, what proportion do they constitute of modern Europeans following the spread of agriculture into Europe during the early Holocene? In addressing the second question there is longstanding debate over the origin of modern European populations. One theory, called the 'Demic Diffusion Model', proposes that early farmers migrated from the Middle East (or Near East), introducing agriculture to local European hunter-gatherers and forming 'hybrid' settlements with these locals (Ammermann & Cavalli-Sforza, 1984). These hybrid populations would therefore possess significant genetic similarity to Middle Eastern populations. An alternative theory suggests that agriculture was spread through ideas and practices, rather than through the movement of people. If so, this would result in a much smaller contribution of Middle Eastern genes.

Genetic mapping of both mtDNA and the Y chromosomes of living Europeans (see Section 3.8) appears to provide an agreed answer to this fundamental question (Underhill *et al.* 2000; Ke *et al.* 2001; Wells *et al.*, 2001; Sykes, 2001). More than 80% of European men have inherited their Y chromosomes from Palaeolithic ancestors who lived in Europe 25 to 40 kya. Only 20% of Europeans trace their Y-chromosome ancestry to Neolithic farmers who moved in from the

Middle East. The data from both genetic lineages not only enabled researchers to trace the movements of the first farmers, they also paint a remarkably detailed picture of the identity and movements of ancient Europeans. The Y-chromosome team, led by geneticists Ornella Semino of the University of Pavia in Italy and fellow researchers from a number of other universities, also took the bold step of explicitly connecting genetic and archaeological data (Semino *et al.*, 2000). They linked two early migrations recorded by the Y chromosome to the Aurignacian and Gravettian cultures.

Using samples from 1007 European men, the Y-chromosome data suggested that most of the men could be sorted into 10 different Y-chromosome variants or haplotypes. The researchers sorted those haplotypes on a phylogenetic tree and used the geographical distributions of modern markers to trace the evolution and spread of the ancient markers. For example, they found that four modern haplotypes, which account for the Y chromosomes of 80% of European men were descended from two now-vanished haplotypes. One, M173, arose more than 40 kya from an even older marker called M45. Apparently M45 was present in men living in Asia at this time, as other descendants of this haplotype are now seen in Siberians and Native Americans. Meanwhile, the descendants of the M173 marker are found at the highest frequency today in Europe. So the researchers concluded that M173 is an ancient Eurasiatic marker that moved into Europe around 40 kya. They have proposed that this marker is the 'signature of the Aurignacian', and that this cultural group came from central Asia. If the team is right, then half of modern European men still carry the genetic signature of this ancient cultural group. Using similar reasoning, the researchers report that the next wave of migration into Europe, marked by a mutation known as M170, occurred about 22 kya from the Middle East. The authors link this wave to the Gravettian culture. The dates of these waves are a bit later than those presented in Section 3.9, but given the uncertainties in the absolute dates obtained from genetic analyses they correspond reasonably well with the archaeological data.

Once in Europe, the timing and geographical distribution of markers suggests that Aurignacian group remained most prevalent in western and southern Europe, while the Gravettian group thrived in eastern and central Europe. As the climate worsened during the LGM, those carrying the 'Aurignacian' marker apparently withdrew to refugia in the Iberian Peninsula and the Ukraine, while the Gravettian culture collected in the Balkans. After the LGM these groups moved out of the refugia and their populations expanded rapidly (see Sections 5.10 and 5.11).

The new mtDNA data tell much the same story, with 80% of European women having the older Palaeolithic markers and 20% having Neolithic markers – although in women, the Neolithic haplotypes are not concentrated along the Mediterranean coastline, a finding that could reflect the different movements of the sexes (Richards *et al.*, 2000). But the mtDNA data also suggest the presence of ice-age refugia in Iberia and, to a lesser extent, southern Europe.

The feature of the new Y-chromosome data is that it reveals the geographical sources of origin of the various lineages more clearly. This is probably because in many societies women moved to join their husbands' families, while related men cluster more closely geographically (Wilson *et al.*, 2001). Furthermore, because some men have many children, they leave more offspring with identical Y chromosomes, which gives a sharper geographical signal.

This apparent agreement has been thrown into question by new research. Lounes Chikhi of University College London and colleagues examined data from a different region on the Y chromosome to uncover the contribution of Middle Eastern genes to the modern European gene pool (Chikhi *et al.*, (2002). This produces a higher figure of 50% for the proportion that Middle Eastern farmers have contributed to the modern European population (see Section 5.11). Nevertheless, although the debate rumbles on, the broad consensus is that about three-quarters of Europeans are descendants of those communities that survived in ice-age refugia on the southern fringes of the North European Plain.

4.4 PHYSIQUE

As modern humans spread out across the globe, their physique and, in particular, physical size changed in response to the environmental and climatic challenges they met. As with everything to do with the evolution of modern humans, measuring physical size is bedevilled by the limitations of the archaeological record. Nevertheless, the available evidence clearly shows that our earliest ancestors were big, if not as impressive as their archaic human cousins the Neanderthals.

Using calibrated data from living humans, measurements of the breadth of the head of the femur can be used to produce figures for the body mass of both Neanderthals and early modern humans (Kappelman, 1997; Ruff *et al.*, 1997). These show that Neanderthals, from the period 75 to 36 kya, were on average 30% larger (at 76 kg) than a representative sample of the present-day humans (58 kg), and 24% larger than humans currently living north of 30° N. Even these figures may underestimate the formidable nature of Neanderthals. With their more strongly constructed skeleton and pronounced muscle development, they may have been appreciably heavier than the figures based on modern humans would suggest.

The figures for early modern humans dating from the period 35 to 21 kya are somewhat lower than Neanderthals, averaging 67 kg, while those from the period 21 to 10 kya average 63 kg. It is not clear why their size appears to have declined over this period. By the beginning of the Holocene the body mass of males living at higher latitudes was not significantly different from that of males living at these latitudes today. The same is also true of high-latitude females. At lower latitudes, however, males and females were still significantly larger (11–12%) than those now living at these latitudes. One other interesting feature of these changes is that at the same time there appears to have been a parallel reduction in brain size.

The other aspect of human physique that requires some comment is the adaptation to heat and cold. The comparison most frequently used to illustrate body form differences in the human thermoregulatory system is that between an African Nilotic and an Arctic Inuit. The first is long-limbed with a lean build. The second is squat with short limbs. The former loses about 16% more energy per unit body mass than the latter. So the Nilotic, who has a greater ratio of surface area to body mass, is better adapted to the demands of the tropics, whereas the Inuit is suited to the Arctic. More generally, there is a highly significant correlation between latitude and absolute body breadth in a large number of modern populations. Nutritional factors are unlikely to explain this geographical variation.

Neanderthals had a relatively wide body and this body morphology, which among modern humans is always associated with cold environments, lends strong support to the view that they were 'cold adapted'. The modern humans that displaced Neanderthals in Europe did not have the same body proportions but, instead, were defined as being 'gracile' (longer and leaner), suggesting a more equatorial origin for these populations with subsequent northward migration into Europe. The changes in the physique of the various groups of people in Eurasia reflect the environmental challenges they faced over the last 40 kyr or so in different parts of the great continent. So climatic factors played a central role in how natural selection led to different physical characteristics.

4.5 THE BROAD SPECTRUM REVOLUTION

One of the great debates about life during the ice age centres on how the diet of humans changed. Archaeological analysis suggests that a major element of Neanderthal diet was medium-to-large mammals. Although there is no direct evidence of plant foods, analysis of stable isotopes of carbon and nitrogen (^{13}C and ^{15}N) in bone collagen provides a direct measure of diet and has been applied to two Neanderthals found in a cave in Croatia (Richards *et al.*, 2001). The isotope evidence overwhelmingly points to the Neanderthals having obtained almost all of their dietary protein from animal sources. Earlier Neanderthals in France and Belgium have yielded similar results.

In contrast, isotope values for remains of modern humans dating to the mid-Upper Paleolithic in Europe indicate significant amounts of aquatic foods (fish, molluscs and/or birds) in some of their diets. Most of this evidence points to greater exploitation of inland freshwater resources. These results are seen as supporting what is known as the 'broad spectrum revolution', which resulted in a broadening of the subsistence economy in inland Europe by the mid-Upper Paleolithic period (Stiner, 2001). One explanation for this shift in diet is that it was associated with significant population increases (see Section 4.6).

Although it is not possible to identify a common thread in this changing diet, what can be said is that farther south, by the end of the LGM, the diet of hunter-gatherers had become surprisingly eclectic. The findings at Ohalo II (see Section 3.13) reveal plant remains from an impressive variety of species (Nadel & Werker,1999). Some 100 species have been identified from the remains of tens of thousands of seeds and fruits. These included acorns, almonds, figs, grapes, olives, pistachios and raspherries plus wild barley and wheat, the ripening times of which covered spring, summer and autumn. Animal bones showed that their prey was principally gazelle, fallow deer, fox and hare, plus many species of birds, and lots of fish.

Another important source of vegetable matter for hunters who preyed off migrating ruminants, such as reindeer, was the stomach contents of the animals they killed. Partially digested lichens are much easier for humans to digest than the alternative of trying to prepare such vegetation in an edible form by cooking, and the result is said to resemble spinach.

The absence of measurements of the plant content of diet in ice age Europe limits what we can say about the diet at the time. We can, however, make some observations about the range of plants that are edible in modern temperate climatic zones. Combining this with analysis of the extent of the knowledge of modern hunter-gatherers of flora and fauna we can safely assume that palaeolithic modern humans living in Europe would have had an exceedingly comprehensive knowledge of the food sources in their local flora. Although these would have been limited in winter they would still have constituted a potentially significant part of their diet.

The range of edible wild plants in modern temperate Europe is extraordinary. It includes berries, bulbs, corms, fruits, fungi, grasses, herbs, leaves, nuts, pulses, rushes and tubers. Many of these are well known to us, such as acorns, beech mast and pine nuts, although the first two, while nutritious, are regarded as too boring for our modern diet, as well as requiring considerable effort to make edible. As for berries, fungi and fruit, the important knowledge was to identify those that were safe to eat: something our ancestors must have accomplished. Others come as a surprise. The fact that the roots of bracken (*Pteridium aquilinum*), various bulrushes and couchgrass (*Agropyron repens*), and the bulbils of the lesser celardine (*Ranunculus ficaria*) are all nutritious sources of carbohydrates is unexpected. Other versatile sources of sustenance include the dandelion (*Taraxacum officinale*), whose young leaves and long taproot are still prized by some rural communities, notably in France, while the flowers make a fabled country wine. The buds, shoots, flowers and berries of the elder (*Sambucus nigra*) are all edible. Similarly, the young leaves of the birch can be eaten, while the sap of both the birch and sycamore, when boiled down, forms tasty syrup. In coastal sites, the combination of abundant shellfish and a variety of edible seaweeds provides the basis for a continuous, if rather unexciting diet. To judge from the vast shell middens found in early Holocene sites around the world, many people must have survived the ice age living close to the sea shore.

The wide range of foods eaten by the people at the end of the LGM was part of the continuing revolution in human behaviour. The widening diet led to other advances including grinding, drying and storing nuts plus exploiting small animals. It has been argued that by adding new species to the diet, humans raised the carrying capacity of an environment constrained by the extreme climatic instability at the end of the LGM. As we have seen in Chapter 2, not only was the short-term variance of the climate exceedingly high, but the incidence of larger changes in the climate also increased at this time. Such factors could have played a significant part in this adaptive process, especially in the shift to greater use of aquatic resources.

Archaeological evidence for broadening of Palaeolithic diets in Eurasia is that there was greater exploitation of energy-rich nuts and large seeds, whose nutritional benefits require considerable work and equipment to extract. This trend is apparent in the increased use of milling tools after the LGM and some evidence of storage facilities and preserved plant parts. In addition, when hunting was poor and times were hard, people would presumably have become less selective about what they ate. The evidence that they ate a wider range of creatures is, however, less clear than that for plants. Indeed, some archaeologists argue that there is no evidence of a broader faunal diet in the later Palaeolithic, although the stable isotope observations quoted at the beginning of the section would suggest otherwise.

The problem with measuring the change in spectrum of game animals is that simple models are insensitive to physical and behavioural differences among prey. The only qualification normally added to these analyses is prey body size, because all game animals are composed of similar tissues, and large animals yield much more food than small ones, even if they are more difficult to catch. This overlooks great differences in the catching of different animals and in the consequences of heavy exploitation of animals that are broadly equivalent in food content: tortoises and shellfish are easy to catch; hares and birds are not. So any effort to establish how changing diet was a measure of changing climate or advancing social organisations has to look closely at the type of prey exploited.

4.6 CONCERNING TORTOISES AND HARES

A different aspect of the diet during the last ice age can be extracted from the archaeological record. Where there is sufficient evidence of the food sources in prehistoric middens, it is possible to extract information about past population fluctuations. To our ancestors of 50 kya, 'fast food' meant just the opposite: it was something slow-moving and easy to catch. Archaeologists can extrapolate from the detritus of ancient meals a measure of prehistoric population size and its growth rate. Results from three caves in north coastal Israel – Kebara, Qafzeh and Hayonim – yield data covering the period from 200 to 10 kya which has been gleaned from the shells and bones found in early human middens (Stiner, 2001). Although this record extends back into the time when Neanderthals were the only humans in the region, the early arrival of modern humans here makes the data from the vicinity of particular interest. The earliest remains showed that slow-moving prey like tortoises and shellfish constituted a greater proportion of the human diet. Population levels remained at a low level until the Upper Palaeolithic. Only then did it start to rise, and with this rise came evidence that the consumption of fast-moving prey like hares and birds began to increase steadily.

A similar set of results emerges from analysis of middens in coastal sites in Liguria and Lazio, in Italy (Stiner, 1999). Here the dominant feature of the diet in the Middle Palaeolithic (dating from 110 to 70 kya) and early and mid-Aurignacian sites (dating from 36 to 27 kya), as revealed from the middens, is shellfish. From the Gravettian sites (28 to 24 kya) the reliance on shellfish drops sharply and the remains are principally of birds, rabbits and hares, as the population rose. The luckless tortoise does not seem to have featured in the diet in this part of the world after the Middle Palaeolithic.

Because capturing hares and birds requires the use of nets, snares and other technological innovations, humans probably did not hunt such creatures until slow prey became hard to find. The growth of fast prey in the human diet over time therefore suggests that groups of hunter-gatherers competing for resources had proliferated to the point that they could no longer simply move on to another site, expecting to find untapped sources of easily captured food. Their subsistence territories would instead begin to infringe on those of other human groups who were competing for the same resources. In short, increasing population might have led to the development of the new trapping technologies.

The second, related conclusion is that the average size of the slow-moving prey became smaller over time. This was a slow process.

Indeed, for 100 kyr, human hunting did not noticeably diminish either the size or the relative abundance of slow prey at the cave sites. At Hayonim half the animal remains from the early Middle Palaeolithic (200 to 100 kya) were tortoises. Only when the pressure on resources crossed a threshold did the tortoises plummet to a mere 7% of the remains. At Kebara, a significant drop in tortoise size did not occur until 30 kya. At the Italian sites, the size of the limpets remained roughly constant until the Gravettian period, and then there was a marked reduction in the size by around 18 kya.

These variations in small game hunting along the northern and eastern rims of the Mediterranean suggest comparatively recent reductions in both size and relative abundance of slow prey in prehistoric diets. This suggests that early human populations were still small and mobile, spending little time foraging in any one area. Results from predator–prey simulation modelling indicate that human population densities increased abruptly during the late Middle Palaeolithic and again after the LGM. These demographic fluctuations, which support the genetic evidence (see Section 4.1), may well have been in response to climatic changes, especially the amelioration after the LGM.

4.7 GENDER ROLES

The extraordinarily demanding conditions of the ice-age Eurasia must have had a profound effect on the roles of men and women. This statement is not made because of any inherently sexist assumption that if hunting played a dominant role in feeding modern humans then the society was implicitly male-oriented. Estimates of the balance of food sources in modern hunter-gatherer societies come up with quite a range of figures. These suggest that, below 40° N, hunting provides between 20 and 50% of the food whereas berries, fruits, leaves, roots, shoots, grubs and eggs provide the remainder. Nevertheless, in the tropics, men hunt while women do the gathering, as the latter is more compatible with the demands of pregnancy, breast-feeding and infant transport. Further north, food gathering is limited by the long winter, providing less than 10% of all food among hunter-gatherers above

60° N at this time of year. The extreme case is the Arctic tundra, where almost all of the available biomass is in the form of game animals. Such environments compel women to process food obtained through hunting instead of gathering food on their own.

There are, however, two fundamental issues to address when extrapolating from these modern analogues to the Upper Palaeolithic. First, the conditions experienced in the present-day Kalahari, or highland New Guinea, are vastly different from those of the ice-age Dordogne or Ukraine. In particular, the survival strategy of living close to the migration route of large herbivores provided a reliable source of prey. It also offered the opportunity of scavenging any animals that fell by the wayside or were killed by other predators. Second, all of us have evolved over the last 20 to 30 kyr. While these changes may seem more obvious in people living in modern cities, the question of just how much of the palaeolithic relic is reflected in modern hunter-gatherer groups is as relevant as asking the same of modern urban dwellers.

The other issue is the whole question of the representations of men and women in the Upper Palaeolithic. Much has been written about the female figurines from this period, often termed 'Venus' figurines with large stomachs and pendulous breasts, which were seen as depicting pregnancy and/or lactation and therefore signified fertility and the successful maintenance of the population. These figurines have been seen either as objects of social control and male desire, which subordinate women's status in society, or, conversely, as a form of 'goddess', which attached the highest cultural importance to female qualities and their biological role. Needless to say, the argument as to what these figurines symbolised to people who lived 25 kya is unresolved.

There is another feature of the Venus figurines that may provide insight into gender roles. This is the fact that some of them are partially clothed (Soffer, Adovasio & Hyland, 2000). The attire depicted includes basket hats or caps, netted snoods, bandeaux, string skirts and belts. These garments were clearly not daywear, but are more likely to be ritual wear, real or imagined, which signified distinct social categories.

This is seen as an indication that spinning, weaving and basket making were female roles, which were highly respected in the community.

Broadening the discussion to the representation of the human form in the Upper Palaeolithic and subsequent prehistoric ages, the debate is clouded by the fact that the obviously female figurines do not all look alike. Furthermore, most figurines from these periods are not female. Indeed, most cannot be ascribed with any certainty to either gender. In addition, so-called 'goddess' figurines form only a small proportion of total number of objects recovered. Then there is considerable doubt as to whether these figurines had any religious significance for their makers. The assumption of religious connotations for prehistoric figures may be a modern construct designed to handle images that did not easily fit into a male-dominated, science-based era. In effect, it was designed to impose upon our prehistoric forebears the same gender roles as held by society in the nineteenth and twentieth centuries. Now most archaeologists would argue that it is better to have an open mind about whether women and men are and have always been the same.

Against this conceptual back drop the first practical thing to emphasise is that in ice-age Eurasia the proportion of the diet met from hunting and scavenging must have been considerable, especially in the winter half of the year. The emerging evidence of diet from the measurement of stable isotopes in bone collagen (see Section 4.5) supports an emphasis on animal protein in the diet. Nevertheless, this does not mean that this was a wholly male activity. The broadening diet in the Upper Palaeolithic to include smaller animals, birds and fish means that women may have played as much of a role as men. The evidence at Dolni Vestonice of weaving and the ability to create nets shows that before the LGM modern humans were capable of trapping small mammals. This was an activity that women could do just as well as men.

The need to avoid falling into obvious stereotypical traps does not mean that the fundamental differences in biological roles can be ignored. The demands of childbearing and child care imposed a heavy burden on women. First and foremost this was reflected in life expectancy. Any attempt, however, to show that gender roles were moulded by life expectancy in the Upper Palaeolithic is fraught with difficulty. The whole subject of palaeodemography has to grapple with the problems of limited archaeological evidence of the human condition. The number of skeletons available is small and there are major challenges with judging the age accurately. For example, a study published in 1961 of 76 Upper Palaeolithic skeletons (Vallois, 1961) found that less than half these individuals had reached the age of 21, that only 12% were over 40, and that not a single woman had reached the age of 30. This final stark statistic suggests that the burdens of childbearing played a crucial part in the low life expectancy of women. A more comprehensive analysis (Angel, 1984), suggested that in the late Upper Palaeolithic men lived on average to around 35 and women to 30. If anything, these figures declined during the early Holocene as people settled down and turned to agriculture, although this may have been a slow process. The transition to agriculture appears, however, to have affected men and women differently. A recent study (see Section 6.5) suggests that, while the mean age at death for men rose, that for women fell, as increased fertility led to higher maternal mortality. This shift could well have had an impact on the gender roles.

The evolutionary implications of any substantial separation of the hunting and gathering functions are profound. Where men were pursuing megafauna there was a premium on strength, notably the ability to throw missiles, explosive bursts of speed and a capacity to overcome fear with bouts of extreme aggression, combined with either the ability to wait lengthy periods for opportunities to ambush their quarry or the patience to track animals over great distances, and the essential need to work as a team. In addition their whole approach to the world around them was governed by understanding the behaviour of the animals they pursued. This involved the anthropomorphisation of their thinking about the behaviour of their quarry. Only by putting themselves into the mind of the animal were they able to anticipate its behaviour. Moreover, in a world of immense climatic variability, the whole process of successfully intercepting passing herds of animals would involve great patience. Conversely, genetically related hyperactive forms of behaviour now considered socially unacceptable, which involve rapidly shifting attention and impulsivity, may well have been of great value in handling the extreme demands of hunting.

The skills needed back at the camp were different. Leaving aside the obvious demand of raising families and the cooperative nature of this activity among women, there was the continual work involved in all aspects of the life. Not for the women the long hours of waiting for their quarry to turn up at a chosen spot, hours filled with idle or nervous banter. Instead talk was part of the process of handling the endless drudgery of much of the work. This would have involved much more personalised discussion of life and a more caring response to the mutual problems they all faced: what we now call gossip. In the highly skilled tasks of locating and collecting food, subtle abilities to recognise patterns and faint changes in shades that provide evidence of hidden resources would have been at a premium. This activity was also enhanced by more communicative forms of talking. Brute physical strength would have counted for less, but there is no reason to think that physical stamina was any less important to one or other gender in surviving the ice age.

A more specific aspect of the ability to find food, where there is a clear gender difference, is colour vision. The gene for colour-blindness is carried on the X chromosome. Women have two copies of the X chromosome. Men have an X and a Y chromosome. So men are more likely to be colour-blind (about 8 per cent of men are red–green colour blind, whilst only about 0.4 per cent of women suffer this disability). In addition, recent work by Brian Verrelli, of Arizona State University, and Sarah Tishkoff, at the University of Maryland (Verrelli & Tishkoff, 2004), found that one of the genes connected to colour vision shows an unusual amount of genetic variation. This variation enhances the ability to discriminate between colours in the red-orange part of the visible spectrum, particularly among females, because of their two copies of the X chromosome. This enhanced red-orange vision may allow females to distinguish better between berries and foliage when they are gathering food. Conversely, it has been argued that in the dappled shade of temperate forests, especially at twilight, red–green colour-blindness may be a benefit in hunting game, but there is no research evidence to back up this assertion.

Clearly, some evolutionary process has been at work in these differences in colour vision. while many hunting and gathering activities would have involved both genders, the separation of the activities for much of the time would have led to fundamental differences in how they viewed and survived the world around them. It is, however, important to emphasise the more general fact that they evolved together. So, while they were bound to be different, they were also complementary. Long before the emergence of hierarchical, male-dominated societies, many of the features needed to respond to the challenges of the ice age for both genders were 'hard-wired' into us. So, deep down we may retain the equality and sharing nauture of the hunter-gatherer societies that enabled our ancestors to survive in a harsh climate.

This brings us to the interesting question of whether the insights we can derive from modern hunter-gatherer societies provide us with a window on life in the ice age. It hinges on the extent to which the challenges facing these remaining groups are comparable to those faced by their forebears. If they are, then we can extrapolate from current observations back into the ice age. But if the conditions during the ice age were appreciably more demanding than the present, then we need to consider what the implications of these differences might be. As has been emphasised frequently, the greater variability of the climate from year to year and sudden large shifts to new conditions must have constituted far greater threats to ice-age people. In extreme cases these changes may have overwhelmed whole communities (see Section 4.1). It follows that these extremes probably served to emphasise certain survival skills of men and women that were less important in the calmer times of the Holocene.

Addressing these questions requires us to return to the limitations of the archaeological record and how this may have been distorted by climatic factors. The dependence on the Eurasian evidence steers us in the direction of emphasising the extreme nature of the annual cycle with its icy winters, brief warm summers and exaggerated fluctuations on the interannual and interdecadal timescale. Exploiting the uncertain migratory movements of big game could well have taken hunting/scavenging parties away from camp for long periods. This lengthy physical separation probably led to a polarisation of gender roles. So, to the extent that Europe, Asia and the Americas are inhabited by the descendants of those who survived the ice age in these rigorous conditions, the presumption of climatic tuning of gender roles may have some justification.

This leaves us with the task of trying to interpolate between the limited bits of evidence from elsewhere. As described in Chapter 3, the evidence from both the Nile Valley and Israel provides a vivid insight into the lifestyles of the people living in these places at the end of the LGM. They are, however, snapshots of life for a short time in a few well-preserved sites. Beyond this there are vast expanses of time and space where there is virtually nothing. What we know about Africa is that it was generally more arid than in the Holocene, but possibly less variable on the shorter timescale. This combination is unlikely to have posed appreciably more demanding conditions for hunter-gatherers, who could move freely over the landscape, than they have experienced during the Holocene (see Section 5.4).

A similar conclusion has to be drawn in respect of the Middle East. Here the obvious feature of the climate during the last ice age was increased aridity. In addition, the more rapid bouts of climate change, plus the increased variability of the climate in mid-latitudes of the northern hemisphere, extended into the Middle East. Nevertheless, the changes from year to year, or decade to decade, posed less of a threat to people living in the region than to those living farther north. This is because the growing season would be much longer and, as the description of the diet at Ohalo II shows (see Section 3.13), this enabled the inhabitants to find food throughout most of the year. Moreover, although there is no archaeological evidence, the living conditions in what is now the Persian Gulf would have been even more manageable, as would many riverside or coastal sites in South Asia.

In summary, if the extremes of the ice-age climate had any implications for gender roles, these were probably most marked for those humans who lived to the north of the mountains from the Pyrenees to the Himalayas, during the period from 40 to 15 kya. For the rest, the fluctuations at lower latitudes would probably have posed the same challenges as have been faced by hunter-gatherer societies during the Holocene (see Chapter 5). So the climatic impact on gender roles here may not have altered appreciably.

4.8 ANTHROPOMORPHISATION: A PATHETIC FALLACY OR THE KEY TO SURVIVAL?

John Ruskin's observation that the attribution of violent feelings to inanimate objects was the 'Pathetic Fallacy' has been extended by some to the anthropomorphisation of flora and fauna. Nevertheless, the human tendency to attach human emotions to animals and even plants is a deeply held propensity. From the earliest age we are exposed to a wide variety of stories and images that show animals as having a range of human characteristics. When extended to the ownership of pets or the husbandry of farm animals this bond develops into strong emotional ties. In a rational world it could be argued that this behaviour is indeed a pathetic fallacy, but the strength of these feelings is clearly a deep part of the human psyche. Its origin may well lie in the vital importance to our ancestors of understanding animals and, in particular, its importance in surviving the ice age. It can also be extended to the deep attachment to nature that could have proved so useful in reading the signs of changing climatic conditions in the world around them.

Modern humans obviously had a profound knowledge of the habits of the many animals they hunted. They set up their camps close to where herds of such mammals as reindeer had to ford streams, when the beasts would have been particularly vulnerable to ambush. Vast accumulations of animal bones, sometimes showing evidence of cooking, have been found in association with stone tools at the ends of blind valleys into which the victims had been driven, or at the bases of cliffs over which they had been stampeded. One such strategy was used until relatively recently in North America to hunt buffalo. It involved one courageous individual, cloaked in the hide of a newly killed baby buffalo, luring the rest of the herd into a brush-fenced defile that led to a precipice. Once in the confines of the fences the rest of the hunters would stampede the buffalo over the cliff. The task of the lure involved not only bravery, but also the ability to imitate the behaviour and the cries of a distressed baby to make the subterfuge effective.

There is one other aspect to imagining oneself in the mind of an animal. This is as a way of overcoming fear. Taking on the character of a fierce beast remains a psychological model for responding to danger. Drawing on the examples of the animals they hunted it is understandable that Palaeolithic hunters could have assumed the belligerence of their prey to steel their resolve. Furthermore, the group example that would have been most natural to adopt is that of the wolves that they may have domesticated at an early stage in the Upper Palaeolithic (see Sections 3.15 and 4.10).

Even more striking is the quality of their cave and rock painting. Clearly, the representation of animals in these works of art was part of a close relationship between humans and the animals around them. The quality of the images from the caves in Europe shows an intimate knowledge of the animals. They provide details of physical features of animals, such as bison in their summer moult, stags baying in the autumn rut, woolly rhinoceroses displaying the skin fold that was visible only in summer, or salmon with the spur on the lower jaw that males develop in the spawning season.

Some of the features of these images might be open to doubt. Because soft-tissue features do not normally survive in the fossil record, we have no archaeological evidence for the fact that the extinct rhinoceroses had long shaggy coats. As for the darkly coloured hump behind the shoulders of the giant deer (also known as the Irish Elk)

(*Megaloceros giganteus*), we have to rely on the record of the picture in the Cougnac Cave in France. The accuracy of these observations is confirmed by the complete specimens of mammoths preserved in the permafrost of Siberia. Right down to their split-tipped trunks, the images are bang on. So we can have confidence in the details of other animals recorded in other images.

The various explanations that have been put forward for the underlying objectives of the people who created these images are summarised in David Lewis-Williams' *The Mind in the Cave* (see Bibliography). The most direct reason would appear to be the intention to control the movements and lives of animals. Associated with this aim to control, there may have been a hope that the images could influence the weather to make hunting more successful. At the more obviously spiritual level the act of creating the images would provide contact with the spirit world and supernatural forces. There may also have been the hope that intercession with the spirit world could heal the sick.

There is another aspect of this ability to interpret and anticipate the behaviour of animals. It implies an understanding of the seasonal changes throughout the year and an ability to plan to intercept the migratory patterns of large animals. In those circumstances where people lived permanently on the migration route, as appears to have been the case at Dolni Vestonice (see Section 3.10) or present-day Barrow, this was relatively simple. If, however, it involved moving from some more permanent settlement to intercept the spring or autumn migration of reindeer, or the early spring run of salmon, then timing was of the essence. When combined with a migratory lifestyle this skill would only work if groups could return to the same spot at the same time of the year.

It is often assumed that it was only with the coming of agriculture that humans needed to know precisely when to make decisions on the timing of, say, planting crops. But long before this, in ice-age Europe, where the variations of the seasons were even greater than in modern times, the ability to keep track of the state of the world around them throughout the annual cycle of life was essential to survival. The greater variability of the weather from year to year made this skill an even more vital part of human existence.

The timing of migratory patterns would not have been restricted to the movement of specific prey. The behaviour of birds in particular could have provided valuable information. This would have relied on knowing the normal time of arrival: a useful marker in a natural calendar. A recent analysis of lengthy records of the migration of birds arriving in and passing through the island of Helgoland in the southeastern North Sea provides the most comprehensive insights into how this behaviour responds to climate change (van Noordwijk, 2003). What this work has shown is that bird movements respond both to local climatic conditions and to wider circulation patterns. Short-distance migrants, flying in from continental Europe, are most affected by local temperatures. The timing of the passage of long-distance migrants, flying in from Africa, is closely linked to the pressure patterns over the Atlantic Ocean, as measured by the index of the North Atlantic Oscillation (NAO: see Section 5.1), rather than local temperatures. The marked rise in spring temperatures over the last 40 years, which is in part a reflection of changes in circulation patterns over the North Atlantic, has had measurable effects of migration behaviour of all the birds studied. Birds have been returning earlier in recent years.

Three hypotheses have been proposed for the earlier passage. First, the moment of leaving Africa has not changed, but refuelling in continental Europe proceeds more quickly, because more food – in the form of insects – is available earlier. This would tally with higher values of the NAO index as this indicates favourable springtime conditions across Europe as whole. The second is that the effect of the NAO is perceptible in Africa, enough to stimulate birds to alter their flight times. Third, the weather in Africa does not hold the key, but the effect of climate change at higher latitudes is what matters. This could occur through natural selection on the bird population, by conferring benefits on the offspring of birds that migrated earlier, and so altering the 'trigger time' for starting migration. While these questions remain exciting areas of investigation for today's ornithologists, they may have been vital signals for humans living in ice-age Europe. The fact that birds respond so promptly to climatic shifts means that they provide an accurate measure of more general changes in environmental conditions. This evolutionary response is not a recent development and would, if anything, have been more vital during the ice age. Modern humans who noted the behaviour of birds might have been able to make more informed decisions about when to anticipate the arrival of migrating herbivores.

Ecological awareness would have extended well beyond the behaviour of animals. It would have involved the whole process of observing the timing of the different stages of vegetation. This remains the subject of active scientific investigation known as phenology. The rules of this discipline, involving recording leaf opening, flowering, fruiting and leaf fall, were compiled by the Swedish naturalist Carolus Linnaeus. In England the best-known record is that maintained by six generations of the Marsham family on their estate near Norwich between 1736 and 1947. This included the leafing of 13 kinds of tree, and the extraordinary variation from year to year confirms modern gardeners' experience of how erratic the progress of spring can be (Sparks & Carey, 1995).

In the case of hawthorn (*Crataegus monogyna*) the earliest leafing date was 27 January in 1804, while the latest was 26 April in 1917. Similarly, the leafing date of the common elm (*Ulmus procera*) varied from the beginning of February to the beginning of May. Other species of native tree typically showed a variation of around two months between the earliest and latest springs. But the timing of various species is a complicated response to different combinations of temperature, rainfall and sunshine. The one obvious feature is that cold wet springs lead to late leafing and flowering while warm, dry ones bring things out early.

The much more difficult aspect to unravel is the balance between the response to the weather and the underlying control of the length of the day. There is a vast array of studies by biologists and horticulturists of how changing lighting and temperature conditions can be used to alter and control the growth characteristics. This work has established that while either increasing or decreasing the amount of light they receive each day can control the flowering characteristics of some plants, others are relatively independent of this control. But there is little scientific evidence about whether daylight or temperature is the main factor in the emergence of trees and shrubs from dormancy. The general conclusion is that in most cases it is a bit of both with some species being more sensitive to daylight and others more controlled by temperature (Burroughs, 1998).

If this is the best modern science can do, just imagine how confusing the more variable conditions of the ice age would have been to the hunter-gatherers of the time. Nevertheless, the skills of modern hunter-gatherers in interpreting the behaviour of flora and fauna are prodigious. So it is reasonable to assume that the fact that humans managed to survive the ice age shows they were practised at the art of interpreting the signs of changing vegetation and patterns of behaviour of wildlife, and whether these were advanced or retarded in respect to the normal seasonal cycle.

All of this adds up to a powerful set of reasons as to why, even today, an inherent empathy for the animals we raise or hunt has been retained as an integral part of our make-up. Refined by natural selection in a world where climate variability kept human existence on a knife-edge, the ability to anticipate and exploit the behaviour of animals was crucial. The transition from hunter-gatherer to agricultural lifestyles may only have served to reinforce this empathy as the domestication of animals involved a similar but subtly different set of skills.

4.9 THE IMPORTANCE OF NETWORKS

Throughout the Middle and Upper Palaeolithic there is evidence of the movement of certain valued goods over long distances. The most obvious example of this network of exchange is in flints. The distances over which these were moved increased at the beginning of the Upper

Palaeolithic. Surveys of the movement of raw materials for stone tools show that between the late Middle Palaeolithic and early Upper Palaeolithic the distribution of the distance of transfer shifted markedly, from virtually none being moved over 100 km to nearly half being exchanged over 200 km or more. In addition, there is evidence of seashells being moved over even greater distances (Gamble, 1999; pp. 312–318). The importance of this emerging activity is not simply a matter of acquiring material goods but is a measure of the increasing sophistication of palaeolithic societies between 40 and 30 kya: the purpose of exchange was to develop social contacts. This raises fascinating questions about the network of contacts that existed between communities and whether this structure provided a mechanism to enable people in Europe to survive the LGM. This may involve the emergence of complex language and other forms of symbolic communication (Mellars, 2004), which may well be the defining characteristics of both the ascendancy of modern humans and the decline of the Neanderthals.

One particular aspect of mobility at this time is the evidence from genetic mapping of the difference in the diversity of mtDNA and Y-chromosome data. With mtDNA there is much more similarity between populations than with Y chromosomes (see Section 4.3) (Seielstad, Minch & Cavalli-Sforza, 1998). This suggests that women had a much higher migration rate than men. One possible explanation is that men often travelled to find wives and returned home with them. Leaving aside all issues of 'sexist' interpretations of what might be regarded as 'trade' in this context, the movement of people as a part of life in the Upper Palaeolithic must be seen as a vital defensive mechanism in increasing the gene pool, sustaining population levels and maintaining cooperative links between vulnerable groups in a hostile environment.

The evidence of more extensive exchange networks may also reflect a migratory lifestyle. This offers incentives for the invention of efficient modes of transportation, of cultural means of anticipating future events, and of methods of communicating information and maintaining social networks over space and time. Furthermore, when practised in areas of seasonally abundant food, such a lifestyle would have permitted major population explosions and propelled social changes. These upsurges are more likely to have occurred during interstadials, which in terms of our climatic template means DO events 8 to 3 (see Fig. 2.6 and Table 2.1), and probably encouraged the invention of art, personal decoration and higher-quality tools.

The eventual result of these innovations would have been to enable modern humans to displace less culturally advanced but otherwise highly intelligent populations. Networks played an essential part in this process. In the case of flints, there is little evidence of Neanderthals, living in flint-poor areas, obtaining higher-quality raw materials from afar. They relied heavily on materials available locally. By comparison, modern humans living in similar localities obtained a far higher proportion from distant flint-rich areas. Even in well-endowed areas, groups exchanged appreciable quantities of material over tens of kilometres. Modern humans networked well, and when times got hard they had kith and kin to run to. The Neanderthals, it appears, did not.

4.10 DID WE DOMESTICATE DOGS OR DID DOGS DOMESTICATE US?

If we were looking for an analogue for how modern Europeans might fare in conditions akin to northern Europe during the ice age, the Klondike gold rush comes to mind. Thousands of people (mainly men) set off from the comfort of the cities of temperate latitudes with a limited set of possessions to find their fortunes in the frozen north. A colourful evocation of these times is found in Jack London's books, such as *White Fang* and *The Call of the Wild*. The most striking aspect of many of these tales is how dogs and wolves are depicted as being much better equipped than men to survive the rigours of the savage Yukon winters. In a fictional way it throws considerable light on the questions raised in Chapter 3 about the genetic evidence that we may have had a symbiotic relationship with dogs during the ice age. While the question in the title to this section is put with tongue in cheek, the underlying thought is interesting. Working with dogs may have played a vital part in the survival of ice-age hunters (see Section 3.15). In particular, during the worst climatic intervals, when survival was on a knife's edge, those groups that exploited this symbiosis most effectively could have increased their chances of coming through. It follows that this natural selection would work in favour of those humans who worked best with dogs.

Paul Taçon, at the Australian Museum, and bio-archaeology consultant Colin Pardoe have explored this idea (Taçon & Pardoe, 2002). They propose that wolves or dogs were instrumental in helping humans to survive and thrive in the harsh ice-age environment. Wolves, unlike primates, are fiercely territorial, which is a trait that may have been adopted by humans. They have a much more acute sense of smell to help alert us to danger and to hunt. They can also assist in hunting and, being the most geographically dispersed of all mammals except humans, their help might have assisted our own ancestors in living in harsher environments. Most important of all, the cooperation evident in wolf packs may have given humans the impetus to cooperate more amongst themselves.

5 Emerging from the ice age

> And the waters prevailed exceedingly upon the earth;
> And all the high hills that were under the whole heaven were covered.
> Fifteen cubits upwards did the waters prevail;
> And the mountains were covered.
>
> *Genesis* 7, 19–20

Defining the meaning of 'emerging from the last ice age' is a matter for debate. Frequently it has been assumed that this coincides with start of the Holocene, broadly speaking around 10 to 11 kya, immediately after the Younger Dryas around 12.9 to 11.6 kya. Many of the most interesting aspects of human development appear, however, to have been stimulated by the changes that occurred following the Last Glacial Maximum (LGM). The warming of the Bølling/Allerød (starting 14.5 kya), even allowing for the temporary setback of the Older Dryas at 14.1 kya, had already led to considerable changes. Moreover, even though the Younger Dryas was a dramatic reversal, especially around the North Atlantic, it led to only a limited pause in the rise in sea level (see Fig. 2.10). Nevertheless, in terms of human activities, it may have constituted a much greater challenge. As such it could represent a defining point in human history. Although the Bølling/Allerød probably marks the end of the ice age, the fact that the Younger Dryas is more important in terms of discussing developments around the world means that it is wise to be flexible about constitutes the end of the ice age.

The same flexibility exists in the archaeological record. The question of the timing of the Mesolithic – a term that is usually used to denote the period between the end of the ice age and the establishment of agriculture, which somewhat confusingly is called the *Epipalaeolithic* in the Middle East – has equally fuzzy boundaries. Here again, while hunter-gathers were already adapting to a warmer climate, the Younger Dryas is the most obvious choice of the defining event. So, for the purposes of this book, the beginning of the Mesolithic is the end of the Younger Dryas. As for the boundary between the end of the Mesolithic and the start of the Neolithic, this is even more difficult to define as it relies on some mixture of the setting up of an agricultural economy, together with some form of hierarchical society and significant impact on the local environment. Given the variation in both space and time of this transition, I propose that when I use the word 'Neolithic', like Humpty Dumpty in *Through the Looking Glass*, 'it means just what I choose it to mean – neither more nor less.'

This flexibility is essential when considering the development of agriculture. This profound change in human society is usually linked with the beginning of the Holocene. The transition between hunter-gatherers and food producers was inevitably a gradual process. There are many reasons why the adoption of agriculture would have run in parallel with earlier lifestyles. As was emphasised in earlier chapters, the shift from an outrageously variable climate to one which allowed an approximate degree of the planning needed for successful agriculture took place around 12 kya. Even if this amelioration had been instantaneous, it would have taken decades if not centuries for it to dawn on people that the patterns of behaviour that had enabled them to gather seeds from wild plants could become more formalised as the climate became more 'predictable'. Furthermore, the adoption of agriculture may have been driven by more fundamental shifts in the climate. So, in looking into the human response to the changes in the climate that occurred in the Middle East after the LGM we need to examine some features of the climate in a postglacial world.

5.1 THE NORTH ATLANTIC OSCILLATION

Much of the analysis in this chapter and the next will focus on events in the Middle East and Europe. How the climate of these parts of the world is affected by global circulation patterns is central to the discussion. In this context, it turns out that the features of the North

Atlantic, which have been so much part of the book up to now, continue to dominate our thinking. In particular, behaviour of the winter circulation that is known as the North Atlantic Oscillation (Marshall *et al.*, 2001) is central to this discussion, notably when interpreting events in the Middle East.

Why should this be so? The reason is that the climate of the Middle East is characterised by cool, wet winters and hot, dry summers. Most of the Middle East lacks access to significant surface and groundwater resources because local evaporation far exceeds precipitation. The one important exception is Turkey, which has abundant precipitation resulting from the orographic capture of winter rainfall from eastward-propagating mid-latitude storms generated in the Atlantic Ocean and the eastern Mediterranean Sea. This is vital to the life-giving qualities of the greatest of all the major river systems of the Middle East: the Tigris–Euphrates.

In modern times the flow of the Tigris–Euphrates has two primary flooding periods (Cullen *et al.*, 2002). The first is rainfall-driven run-off from December through March, regulated on interannual to decadal timescales by the NAO as reflected in local precipitation and temperature. The second period, from April to June, reflects spring snowmelt and contributes over half of annual run-off. At the most basic level, this is related to the amount of snow that falls during the winter and the weather conditions during the thaw.

The importance of the NAO is that it is intimately linked to the climate of Greenland and northern Europe. Throughout the preceding chapters reliance on the detailed records of changes in the weather in Greenland have been used to make inferences about changes in other parts of the northern hemisphere. Because we were dealing with relatively large changes in the climate, it was often relatively easy to find examples of proxy records that confirmed that events recorded in either Greenland or in North Atlantic sediments were identifiable across much of Eurasia. As the changes become smaller, this correspondence is more difficult to establish. So we now need to look at more subtle features of Holocene weather patterns in the northern hemisphere. In this respect, the NAO is a particular useful measure of the circulation, especially in winter.

Since the eighteenth century it has been known that when winters are unusually warm in western Greenland, they are severe in northern Europe and vice versa (Van Loon & Rogers, 1978). As the missionary Hans Egede Saabye observed in a diary he kept during the period 1770–78: 'In Greenland all winters are severe, yet they are not alike. The Danes have noted that when the winter in Denmark was severe, as we perceive it, the winter in Greenland in its manner was mild, and conversely.' This seesaw behaviour was quantified by Sir Gilbert Walker in the 1920s in terms of pressure differences between Iceland and southern Europe and defined as the NAO (Walker, 1928).

The NAO shifts between a deep depression near Iceland and high pressure around the Azores, which produces strong westerly winds, and the reverse pattern with much weaker circulation. The strong westerly pattern pushes mild air across Europe and into Russia, while pulling cold air southwards over western Greenland. This flow also pulls cold air down into the eastern Mediterranean and Middle East producing colder, wetter winters than normal. The strong westerly flow also tends to bring mild winters to most of North America. One significant climatic effect is the reduction of snow cover, not only during the winter, but also well into the spring. The reverse meandering pattern often features a blocking anticyclone over Iceland or Scandinavia, which pulls arctic air down into Europe, with mild air being funnelled up towards Greenland. This produces much more extensive continental snow cover, which reinforces the cold weather in Scandinavia and eastern Europe, and this often extends well into spring while the abnormal snow remains in place.

Changes in sea-ice cover in both the Labrador and Greenland Seas as well as over the Arctic appear to be well correlated with the NAO, and the relationship between the sea-level pressure and ice anomaly fields suggests that atmospheric circulation patterns force the sea ice variations (Deser, Walsh & Timlin, 1999). Feedbacks or other influences of winter ice anomalies on the atmosphere have been more difficult to detect. But the frequency of depressions appears to have increased and atmospheric pressure decreased where ice margins have retreated, although these changes differ from those directly associated with the NAO. It may even be possible that a period of one phase of the oscillation produces the right combination of patterns of sea surface temperatures and deep-water production eventually to switch it into the opposite phase.

Understanding the NAO is central to interpreting modern climatic events, because of the influence it exerts on average temperatures in the northern hemisphere. Of all seasons, winters show the greatest variance, and so annual temperatures tend to be heavily influenced by whether the winter was very mild or very cold. When the NAO is in its strong westerly phase, its benign impact over much of northern Eurasia and North America outweighs the cooling around Greenland, and this shows up in the annual figures. A significant part of the global warming since the mid-1980s has thus been associated with the very mild winters in the northern hemisphere (Hurrell, 1995). Indeed, since 1935 the NAO on its own can explain nearly a third of the variance in winter temperatures for the latitudes 20° to 90° N.

These modern observations can help us exploit the evidence of climate fluctuations in Greenland during the Holocene. What the NAO shows is that there are close links between events all around the northern hemisphere even when the climate is in a more quiescent state. So it is possible to extend our thinking beyond the transparently global fluctuations of the last ice age when it is safe to assume that as the climate changed rapidly in Greenland, Eurasia and North America were having their fair share of climatic upheaval. In the Holocene, when the connections are subtler, the lessons of the NAO are more important as a strong westerly circulation over the North Atlantic will lead to a consistent pattern of above and below normal precipitation and temperatures around the northern hemisphere. Conversely, weaker circulation will tend to produce the reverse pattern. The value of this correlation is that it enables us to use sometimes -fragmentary data from around the northern hemisphere to build up a more coherent picture of regional climate change.

As was noted in Chapter 2, there have been at least four global periods of rapid climate change during the Holocene. The first of these was the period around 9 to 8 kya, with a notable cold spell around 8.2 kya in the North Atlantic. Then there was widespread rapid change between 6 and 5 kya, plus periods between 3.5 and 2.5 kya and since 0.6 kya. These more turbulent periods show up in the form of stronger mid-latitude circulation in the northern hemisphere, expansion of mountain glaciers around the world and greater ice formation in the northern North Atlantic, as seen in the amount of rafted debris found in ocean sediments.

The modern observations of the NAO show that when the circulation is in a strong westerly phase, winters are warmer and wetter over Iceland, the British Isles and Scandinavia. Farther south there is a reduction in rainfall in a band from the Azores to the Black Sea, and colder winters in the Eastern Mediterranean. In summer, northern Europe is cooler and wetter, which explains glacier expansion in the Alps and Scandinavia at such times. Conversely, when the circulation is more meridional, the pattern of temperature and rainfall anomalies reverses. In particular, there is increased winter rainfall over the Iberian Peninsula, Italy, the Balkans and Anatolia. This means that the periods of more intense circulation in the North Atlantic during the Holocene would have made life more difficult for Neolithic communities north of the Alps. On the other hand, those living around the Mediterranean and in Anatolia and the Middle East may have benefited from the increased winter rainfall.

This use of the NAO to interpret events during the Holocene has to be qualified by noting that, even when the climate had warmed up dramatically, it took a long time for the system to settle into what we would regard as recognisably modern climatic patterns. This may be of particular relevance to the 200-year cold spell that struck around 8.2 kya. This event was more nearly a throwback to the last ice age for two reasons. First, the remnants of the great ice sheets, most noticeably over Canada, still exerted a considerable influence on hemispheric weather patterns. Second, it is generally accepted that this event was a consequence of the massive release of meltwaters from Lake Agassiz via the Hudson Bay (Barber *et al.*, 1999). This probably disrupted the thermohaline circulation of the North Atlantic for the duration of the event. So the conditions at this time were sufficiently different from those in recent centuries to require considerable care in using the modern NAO analogue to interpret circulation patterns at 8.2 kya.

The shifts in the climate around 4.2 kya are also difficult to interpret. In this instance, the palaeoclimatic data are equivocal. The Greenland ice cores and North Atlantic sediment measurements suggest that around this time there was a short and dramatic decline in the extent of sea ice. This appears to have been linked to a period of weaker circulation in mid-latitudes, except across North America, and in Europe glaciers either retreated or remained stable. There was also widespread drought in equatorial Africa and the Middle East. Although this event was not as extensive as others during the Holocene, it was sufficient to have a significant impact on early civilisations.

5.2 EUROPE, THE MIDDLE EAST AND NORTH AFRICA

The changes in the climate across northern Europe after the LGM reflected closely the records presented in Chapter 2 for Greenland and the North Atlantic. So the succession of the Bølling, Older Dryas and Allerød, followed by the clear setback of the Younger Dryas, can all clearly be seen in the proxy records of climate change across the continent (see Fig. 2.8). Analyses of beetle assemblages (Coope *et al.*, 1998) from Ireland in the west to Finland and Poland in the east have confirmed the earlier pollen analyses and provided estimates of July temperature levels. As the dying throes of the ice age swung the temperature and precipitation patterns north and south, the flora and fauna responded first to the relaxation of the glacial conditions and then had to retreat in the face of renewed cold.

During the Younger Dryas there was a temporary disappearance of the woodland cover that had previously extended over much of Europe and a replacement by dry steppe and steppe-tundra. Across northwestern Europe, the conditions may have been less severe, with forest-steppe (a mixture of patches of trees and grassland) being widespread. Indeed there may have been forest tundra, intermingled with some steppe elements, across most of Poland and Germany, but close to the Fennoscardian ice sheet there was shrub tundra.

In southeastern Europe and the Levant, with the warming after the LGM, precipitation rose to a peak around 13.5 kya. Thereafter, the aridity during the Younger Dryas may have been much more severe than farther north. Pollen analysis suggests that in many areas of Greece and across Turkey, it was even more arid than during the most extreme part of the LGM. Annual precipitation may have been less than 150 mm across much of lowland Greece, and this aridity extended into northwest Syria, Turkey and the western Zagros mountains of Iran, but Northern Israel was perhaps less arid than other parts of the region. Temperatures were also markedly lower.

When the Younger Dryas ended, the change for northern Europe was dramatic, even allowing for the occasional hiccup like the Preboreal Oscillation. In the British Isles the combined influence of the ice sheet over Scotland and the ice cover over the North Atlantic to the west had led to particularly low temperatures. Here, annual temperatures rose by about 15 °C, with a midsummer rise of about 5 °C, and in midwinter by over 20 °C. Comparably large changes occurred across most of northern Europe, although the places closest to the ice sheets over northern Britain and Scandinavia remained under the baleful influence of their icy neighbours.

The succession of changes in the Holocene climate in Europe was initially identified in terms of pollen data (see Section 2.6), which apply to most of northern Europe. This sequence recognises five general climate periods during the Holocene: the *Preboreal* and *Boreal* (11.5 to 9 kya, a rapid transition period followed by a warm and dry period that reflected the more continental nature of the early Holocene in Europe) with summers warmer than now but colder winters than at present; the *Atlantic* (9 to 6 kya, a warm and wet period); the *Sub-Boreal* (6 to 2.5 kya, a warm and dry period) and the *Sub-Atlantic* (2.5 kya to the present, a cool and wet period).

A recent and much more comprehensive analysis of pollen data examined well over 2000 records from North Africa and Europe west of the Urals (Davis *et al.*, 2003). This has produced estimates of changes in temperature for six regions of Europe (covering the east and west halves of the continent for northerly, central and southern latitude zones) since 12 kya. This work confirms the broad features of the earlier analysis described above for the northern and central part of the continent. The interesting clarification of these results is that they show that most of the changes in temperature, notably the warming in the mid-Holocene, occurred in the summer. By comparison winter temperatures have changed little since 8 kya, with western Europe, if anything, showing a continuing warming trend up until recent times.

The analysis does, however, give a different picture for southern Europe. The striking feature is how much the western Mediterranean has warmed since around 8 kya. Here the temperature has risen by about 2 °C in both summer and winter. In the eastern Mediterranean the rise has been much less and largely restricted to winter. More interesting is the decline in temperature here in the period from 11 to 8 kya during which the winter temperature dropped by over 4 °C and in summer by nearly 2 °C, in both cases from levels some 2 °C above current values. In the western Mediterranean the drop in temperature was restricted to summertime, and was less dramatic, falling from around 0.5 °C below current values at 10 kya to 2 °C below current values at 8 kya. The explanation for these changes was the onset of markedly wetter conditions throughout the Mediterranean in the early Holocene.

To identify the impact of specific events, such as the cooling at 8.2 kya, we have to turn to other records, as their impact is not easily picked out in the pollen records. In Greenland it led to a cooling of about 2.7 °C. Analysis of the isotope ratio of the shells of small crustaceans (*ostracod valves*) preserved in the sediment of Ammersee, a small deep lake in southern Germany (see Fig. 2.8), suggests a cooling of about 1.7 °C in the annual air temperature at the time (Von Grafenstein *et al.*, 1998). Studies in Lake Annecy, in nearby eastern France, indicate an increase in precipitation at the time (Magny *et al.*, 2003). This work also concludes on the basis of these data, together with other hydrological records produced in Europe for the same period, that mid-latitudes between around 50 and 43° N experienced wetter conditions during this cooling episode, whereas northern and southern Europe and the Mediterranean had a drier climate. These wetter conditions appear to have extended eastwards as far as Lake Van in eastern Anatolia (Wick, Lemcke & Sturm, 2003).

The decline in rainfall in the Mediterranean appears to have extended to the Arabian Sea. Ocean-sediment records here show a marked decline in the flow of freshwater from the Indus River (Staubwasser *et al.*, 2003), while speleothem records from Oman confirm that the summer monsoon weakened when temperatures were low over the North Atlantic (Fleitmann *et al.*, 2003). In equatorial Africa records of lake levels show a sharp decline suggesting a weakening of the monsoon rains to the north of the Equator (Stager & Mayewski 1997; Gasse, 2000). So it is reasonable to assume that when looking for evidence of climatic-induced change in human affairs, any events that coincide with 8.2 kya deserve close scrutiny. But, as noted in Section 5.1, we need to be careful about using present NAO behaviour to interpret its global impact.

Cores from the eastern Mediterranean and Aegean Sea provide additional information (Rohling, 2002). There are three markedly cooler interludes centred on 8.2, 6.2 and 3.2 kya. Rainfall levels rose markedly at the beginning of the Holocene (10.5 to 9.5 kya) and then drifted back down to levels more typical of modern times by around 5 kya. Possibly more intriguing are the overall changes that appear to have altered the circulation of the Mediterranean around 9.5 kya. Starting around this time oxygen-starvation (*anoxia*) wiped out the entire deep ecosystem of the eastern Mediterranean. This condition prevailed for some 3.5 kyr. Environmental evidence of such anoxic episodes is found in dark, olive green to pitch black, organic-rich sedimentary layers (*sapropels*). What this shows is that the amount of freshwater from both rainfall and river run-off increased substantially, reducing the salinity (in the eastern Mediterranean the water is currently very saline and sufficiently dense to sink to great depths). The less dense layer formed a cap and switched off the formation of oxygenated deep water.

Although this episode was a catastrophe for the deep ecosystem, for people living in parts of the region it was a peculiarly benign climate. The extension of monsoon rains from the south covered much of the Sahara with savannah vegetation that suited nomadic herdsmen. In the Nile valley the effect was for the much greater stream flow to scour deep into the thick sediment, which had formed during the period since the end of the LGM, making the valley less habitable.

5.3 EAST AND SOUTH ASIA

Studies of the climate of China after the LGM rely heavily on analysis of layers of wind-blown loess laid down during and after the ice age. This loess, which is a feature of much of the northern lowlands of the country, is difficult to date. The broad conclusion that has emerged is, however, that the Younger Dryas climate in China was primarily cold and dry, with a brief interruption of warmer, wetter conditions. Overall, it can be assumed that this cold period had an impact comparable to that seen in Europe. As the climate warmed up in the early Holocene the summer monsoon spread northwards reaching levels comparable to current conditions around 9 kya. This expansion continued into the mid-Holocene.

By 8 kya, the south Asian region in general seems to have been strikingly moister and slightly warmer than at present. The increased moistness fits in with a general pattern extending all the way across the subtropics to northern Africa, reflecting greater summer monsoon rainfall at that time. Widespread evidence of lake levels in China and Mongolia shows that conditions were moister than present until 5 kya. In the loess plateau area of north-central China the proportion of drought-tolerant vegetation seems to have reached its lowest point during the early-to-mid-Holocene, before increasing slightly towards the present. Many forest tree species grew further north and west between 8 and 5 kya than at present, indicating precipitation about 100 mm higher than today in many areas of China, and temperatures were about 2 to 4 °C warmer. The deciduous forest of northeastern China also extended 200–300 km northwards into the Russian Far East.

Results from Indonesia shed additional light on the cooling event 8.2 kya. Measurements made of *Porites* corals found within an uplifted palaeo-reef in Alor, Indonesia, dating from 8.4 to 7.6 kya show that sea surface temperatures were essentially the same as today for the first 300 years (Gagan *et al.*, 2002). Then there was an abrupt ~3 °C cooling over a period of ~100 years, reaching a minimum around 8 kya, before returning to current values around 7.6 kya. The cooling was rapid and nearly synchronous with abrupt cooling in the North Atlantic region. This supports the hypothesis that abrupt climate change at high latitudes around 8.2 kya propagated rapidly to the tropics.

Records from the Arabian Sea and dried up lakes in the Thar Desert (Enzel *et al.*, 1999) indicate an increase in the southwestern monsoon activity around 10 kya. Although there is some evidence from lakes in Tibet and Rajasthan of a major dry phase somewhere between around 8 and 7 kya, these records do not show the monsoon weakening until about 5.5 kya. But the Thar data suggest that the rainfall input was not simply due to a strengthening of the southwest monsoon. There appears to have been additional winter precipitation, which was an important source of water for maintaining lake levels. The winter precipitation exceeded present values by about 200–300 mm. Annual figures of up to 500 mm greater than present can be inferred from plant fossils and the molluscan fauna. Overall there was a savannah environment for Rajasthan at that time, in contrast to the present semi-desert. Vertebrate fossil evidence from the lower Indus may even imply that close to the river there was a rainforest or rainforest-savannah environment during the mid-Holocene.

5.4 AFRICA AND THE SOUTHERN HEMISPHERE

In considering the impact of climate change on humans, we have concentrated on the records for the mid-latitude northern hemisphere. In terms of human existence there is, however, one major omission: Africa. In part, this gap reflects the fact that we have little evidence of the fluctuations in the climate across Africa throughout the ice age. The main feature of conditions across the continent at this time was increased aridity. At the same time the archaeological evidence of human life in sub-Saharan Africa is sparse. Now we need to do something to redress the balance.

There is an additional reason for looking more closely at Africa after the LGM. This is that for much of our analysis of the northern hemisphere, it has proved reasonable to emphasise a close link between what was happening in Greenland and the rest of the hemisphere. This is not the case for rainfall over tropical Africa, where, to understand the LGM, we have to shift our attention to events in the southern hemisphere, and even as far as Antarctica. At the same time, this will require us to look more closely at the global connections between the Arctic and the Antarctic.

As noted in Chapter 3, during the LGM much of northern Africa was, as now, arid desert. If anything this desert was even more extensive as the tropical rainforests were greatly reduced in extent and the Hadley Cell circulation that fuels the intertropical convergence zone (ICTZ) was considerably weaker. These conditions show up in a high level of wind-blown mineral dust in the sediment cores from the northern tropical Atlantic (see Fig. 2.11), and provide a clear picture of rainfall changes across northern Africa (deMenocal *et al.*, 2000). They show a sudden switch from dry to moister conditions around 14 kya that, with the exception of one or two short interruptions (see below), lasted until around 5 kya. At the same time sediment data from the southern oceans (Hodell *et al.*, 2001) shows that the extent of sea ice receded dramatically around 11.5 kya and there was an early Antarctic Holocene optimum between 10.5 and 9 kya. A cooling started about 6.5 kya, and by around 5.1 kya sea ice had extended northwards to around the extent seen in modern times. What is particularly interesting is that these relatively rapid changes appear to reflect a response to the lengthy cycles associated with the variations in the Earth's orbit that have driven the 100-kyr cycle in the ice ages over the last million years or so (Section 2.11).

Analysis of lake levels in tropical Africa confirms also that there were a number of dry spells (Gasse, 2000). These included the Younger Dryas, and the 8.2-kya and 4.2-kya events, all of which produced notably low water levels. These declines in rainfall were particularly marked in the northern monsoon domain. In the Ethiopian highlands the decline around 4.2 kya had profound effects on the Nile flood (see Section 6.6.).

Comparison of records of events in the North and South Atlantic indicates that climate change in the middle Holocene may have been initiated in the tropics and subtropics and then exported to high latitudes in both hemispheres. The climate changes near 5 kya occurred at approximately the same time as the establishment of modern frequencies of the El Nino Southern Oscillation (ENSO; see Section 5.17) (Rodbell *et al.*, 1999; Sandweiss *et al.*, 2001). Therefore, non-linear ocean–atmosphere interactions in the tropical Pacific may have amplified the consequences of orbital forcing. A strong relationship exists in modern times between the El Nino Southern Oscillation and Antarctic sea ice extent, and this teleconnection may have also been active on longer timescales.

5.5 NORTH AMERICA

The climate over North America after the LGM was governed by the hemispheric changes that were occurring in the climate, the massive presence of the melting Laurentide ice sheet and the fluctuations in the extent of Lake Agassiz. During the Younger Dryas the High Plains returned to cool wet glacial conditions reminiscent of conditions during the LGM. Then, despite the marked warming, the continued presence of an anticyclone over the ice sheet and Lake Agassiz pushed the weather systems southwards. So cool wet conditions prevailed on the northern portions of the Great Plains, the High Plains and the eastern slopes of the adjacent Rocky Mountains. In contrast, Wyoming and Idaho received less moisture, but to the south, the lack of glacial influence increased the monsoonal input from the Pacific (Schuman *et al.*, 2002). The warming between around 10 kya and 8.2 kya led to increased precipitation from the Gulf of Mexico in the summer half of the year (Lovvorn, Frison & Tieszen, 2001). At the same time, across the centre of the continent there was an expansion of prairie eastwards suggesting increased aridity as the glacial anticyclone maintained cooler conditions.

Following the catastrophic draining of Lake Agassiz through the Hudson Strait around 8.2 kya the High Plains became cold, dry and windy, because of loss of northern inputs of moisture from the weakened glacial anticyclone, and because of northward retreat of the polar front, which allowed northward expansion of westerly winds. These conditions lasted about 200 years. Then, with the declining influence of the last remnants of the Laurentide ice sheet, the polar front shifted farther northwards. Because of high summer insolation, temperatures not only rebounded but also increased to an estimated 2.1 °C warmer than present. This rapid rebound in temperature was accompanied by low rainfall, as the circulation in the North Atlantic remained disturbed by the meltwater event of 8.2 kya. There were strong dry westerly summer winds from the Pacific, and lack of precipitation from either the north or south. This created intensely dry, hot and windy conditions across the Mid-continent, often referred to as the 'Altithermal' drought.

The North Atlantic circulation returned to normal by around 7.8 kya. This led to a northward expansion of the monsoonal rains to the High Plains even though summer temperatures remained about

2.5 °C above present values until around 6 kya. The northern border of monsoonal precipitation remained to the north of its current position until around 5 kya. Thereafter rainfall declined and for much of the rest of the Holocene severe drought was more common across the northern plains of the United States than in recent centuries.

The complementary issue of climate changes in Central and South America is intimately bound up with question of changes in ENSO during the Holocene. For this reason this part of the climate review will be covered in Section 5.17.

5.6 MASS EXTINCTIONS OF BIG GAME

One of the greatest unsolved puzzles of the end of the ice age is what killed off the woolly mammoths and many of their large herbivorous companions. It is generally assumed that the mass extinctions of various forms of big game, which are seen as so much a part of the end of the ice age, were restricted to this time of dramatic climate change. For Eurasia this appears to be the correct interpretation of events. But elsewhere the arrival of humans on the scene may have been a more important factor.

The majority of species of large land mammals, reptiles and birds (those weighing more than 45 kilograms, referred to as *megafauna*), in Australia and North America, became extinct over a period of a few thousand years (Dayton, 2001; Miller *et al.*, 1999). This extinction has usually been attributed to the arrival, expansion and migration of human populations who hunted the megafauna, but it has been difficult to eliminate other possible causes, such as climate change. Across Eurasia the disappearance of many large animals seems to have been a more protracted business, although much of the damage appears to have been done at the end of the ice age. Only in Africa did megafauna survive the transition to the Holocene without substantial losses.

Certain recent studies suggest that anthropogenic overkill was the primary cause in Australia and North America. All Australian land mammals, reptiles and birds weighing more than 100 kg, and six of the seven genera with a body mass of 45 to 100 kg, perished during the last ice age. Measurements of burial ages suggest extinction across the continent between 51 and 40 kya, after the arrival of humans, but well before the extreme aridity of the LGM (Roberts *et al.*, 2001). This extinction in Australia occurred tens of millennia before similar events in North and South America, Madagascar and New Zealand, each of which was preceded by the arrival of humans. While there remains a lively debate about the extent to which climate change played a part in this extinction, the case for human involvement is strong.

In North America, the rapid extinction of many large mammals around 12 kya – a time of rapid climate change – has been attributed both to climatic and human factors. A computer model shows that given simple assumptions about human and prey species distributions and ecology, hunting can cause a major mass extinction with surprising rapidity (Alroy, 2001). The model, which combines population dynamics, ecology, conservation and anthropology, shows that humans even at low densities are capable of precipitating the collapse of prey populations. What this hypothesis does not explain is why certain species disappeared while others survived. Why did the mammoth, camel, horse and ground sloth die out, whereas bison, elk and moose survived? Similarly, among the large predators why did the sabre-tooth cat, American lion and short-faced bear disappear whereas the wolf, cougar and grizzly bear survived? This new analysis from Australia and North America seems to support the humans' role in these mass extinctions, but it does not provide a complete answer.

This having been said, the case for climate change being the culprit looks no more convincing. There is a fundamental weakness in the argument that climate change alone provides an explanation. The succession of glacial and interglacial cycles (see Fig. 2.5) had been going on for at least a million years. The fossil record does not show any clear evidence of these fluctuations leading to an increased rate of speciation. (Barnosky *et al.*, 2004a). So it is not reasonable to say that the last ice age was something out of the ordinary. In fact, if anything it appears to have a little less traumatic than its predecessor. Put simply, the big difference at the end of the last ice age is that modern humans were on the scene, and it is probably not going too far to suggest that our species was at the scene of a crime. But the precise timing of human intervention and climatic changes played a major part in the nature of the extinction process (Barnosky *et al.*, 2004b). So, as with so many other aspects of the complex links between archaeology, anthropology and climate change, there remains a heated debate as to the relative contribution of human activities and climatic factors.

The nature of extinctions in Eurasia is less easy to establish. Clearly, the exploitation of the largest species (e.g. mammoth, woolly rhinoceros and giant deer) played a part in their sudden demise around the end of the Younger Dryas (Orlova *et al.*, 2001). A complication in this picture of sudden decline is the fact that there is evidence of lengthy survivals into the Holocene. For instance, recent studies show that, in western Siberia, the giant deer (*Megaloceros giganteus*) survived until around 7.7 kya (Stuart *et al.*, 2004). More striking is that on Wrangel Island, nearly 50 kilometres north of the Arctic Circle between the Chukchi and East Siberian seas, the woolly mammoth lived until around 4 kya, some 6 kyr into the Holocene (Vartanyan, Garutt & Sher, 1993). While the final survivors were diminutive cousins of their mighty forebears, they appeared to have had little problem surviving in the Holocene. This suggests that climate change was not an important factor in their survival and that they were all right until eventually humans caught up with them in their arctic refuge.

The overhunting occurred wherever humans became well established. In the Near East the hunting of gazelle seems to have been particularly effective. At Abu Hureyra the early occupation is marked by a huge number of gazelle bones, which constitute some 80% of the bones found. There is local evidence of rock enclosures that could have been used to drive herds of gazelle into a restricted area where they were all killed. The same approach appears to have been used by modern humans in ice-age Europe, and there is even evidence that, long before this, Neanderthals stampeded mammoths over cliffs.

A related technique was also practised by Native Americans in the form of 'buffalo jumps'. Using a combination of brushwood stockades leading to a precipitous drop they drove buffalo into the trap and then stampeded them over the cliff to their deaths (see Section 4.8). In the Near East the heavy predation associated with this hunting strategy may well have driven gazelle close to extinction and led to the eventual domestication of goats and sheep. Studies of the age of slaughter suggest that goats (Zeder & Hesse, 2000) were first domesticated in the Zagros Mountains of western Iran around 10 kya. Around this time in some settlements there was a shift to the majority of bones being those of young males. This implies a move to a selective slaughter policy that is a sure sign of domestication.

An interesting alternative hypothesis concerning mass extinctions has been put forward by Ross D. E. MacPhee, curator of Mammalogy at the American Museum of Natural History in New York City. An extremely lethal disease, brought over by humans and their dogs when they arrived in Beringia, may have wiped out much of the megafauna (MacPhee & Marx, 1999). He cites examples of more recent extinctions in Hawaii and Australia that were caused by human disease: for instance, Hawaiian honeycreepers, several species of which have become extinct within the past 100 years and the surviving populations of which all live at high altitude. The reason appears to be that the survivors live at altitudes to which avian malaria-carrying mosquitoes cannot go. These mosquitoes were introduced from tropical North America in the mid nineteenth century, probably in the freshwater bilges of ships. The mosquitoes escaped and started biting the native birds, infecting them with malaria, and the birds died in droves. Megafaunal extinctions followed the arrival of humans and their hangers-on (whether dogs, lice or mosquitoes) in Australia, New Guinea, the West Indies and Madagascar, too. The same pattern does not apply to Africa and southern Eurasia, however, as the local fauna had evolved with humans. Efforts to isolate and identify viruses or their antibodies in the bones of mammals that became extinct at the end of the ice age have, so far, failed to find any evidence of killer diseases. Furthermore, opponents of this hypothesis argue that it is unlikely that a disease could have had such a devastating impact on so many different species.

5.7 THE ORIGINS OF AGRICULTURE

The origins of agriculture from yet another area of fierce debate. At issue are not only the role of climate change, but also population growth and the morphology of plants or animals (e.g. the species exploited, site location and the relative importance of animals and plants in the food supply). What is not at issue is the main advantage of agriculture: it can support more people per unit area than living off wild resources. In addition, by staying in one place it is possible to gather more material possessions. The disadvantages are considerable. They include increased labour costs, reduction in the diversity of the diet, reduced mobility and less sharing of possessions. It establishes a less egalitarian social structure. In addition, living closer together and domesticating animals brings more disease. So the obvious question is why did hunter-gatherers take up food production? The most acceptable reason is that circumstances somehow forced them into agriculture, but there are starkly different opinions about what were the driving forces in this process.

Before embarking on this discussion, it is useful to define what we mean by agriculture. Many hunter-gatherer groups exploited the seeds and fruits from wild stands of plants. In so doing they developed means of storing what they had gathered to last them through times when food was scarce. The recent evidence from Ohalo II (see Section 3.13) suggests that this approach had been practised as early as 23 kya. This could have involved some degree of managing the resources by, say, removing weeds to enable the food plants to thrive. This first step along the road to agriculture (*proto-agriculture*) appears to have a long history. To make the important transition of starting to sow some of the harvest to cultivate a bigger crop in the subsequent year required settled conditions. It would be some 10 kyr before this happened. Even more important was the subsequent step that involved selecting those seeds that show improved properties, in terms of both yield and ease of harvesting. Over time this led to *domestication* of specific plants. In terms of archaeological evidence it is the presence of grains showing signs of having undergone this domestication process that is the crucial measure of the adoption of agriculture.

The history of the debate about the origins of agriculture usually starts with V. Gordon Childe, who developed the idea of the 'Neolithic Revolution'. He was a renowned synthesiser of archaeological thinking, who became Professor of Prehistory at Edinburgh University in 1927, and published two highly influential books: *Man Makes Himself* in 1936, and *What Happened in History* in 1942. He argued that agriculture arose as a consequence of abrupt climate change after the end of the last ice age. He proposed that this led to progressive desiccation that forced the withdrawal of humans, animals and plants to the banks of rivers and oases. The close contact that now prevailed between humans, plants and animals thus led to the first attempts at domestication. Where successful, this resulted in rapid population growth and the establishment of permanent settlements. His analysis took a rather simple view of how this challenge developed. Childe argued that the climate shifted from the cool wet conditions at the end of the LGM to the hotter, drier weather of the Holocene. This simplified approach does not tally with the changes that have been presented earlier. Moreover, it does not include a detailed interpretation of the impact of the complicated nature of this transition, with events like the Younger Dryas exerting particular influence over the move to agriculture.

The somewhat sweeping nature of Childe's 'catastrophic' explanation of the origin of agriculture ran into considerable resistance. Instead, a consensus formed around agriculture being the product of 'population pressure'. Around 12 kya hunter-gatherers began to produce significantly more offspring than they could feed. While population levels of hunter-gatherers are generally maintained at a 'carrying capacity', for whatever reason some palaeolithic populations began to grow. This increase led to an inevitable limitation of resources and made the adoption of the hardship and toil of agriculture inevitable. What was not clear was what drove the population explosion: why did hunter-gatherer population dynamics suddenly require people to embrace food production with all its laborious and time-consuming drawbacks rather than sticking to the freedom of hunting and gathering? Part of the story may be that sedentism and the more frequent use of settlements made child rearing easier. Women did not have to carry young children as part of a nomadic existence and this could have led to larger families. Even so, what was missing was some driving force behind the change, and needless to say, climate change seems to be an obvious possibility.

The two most important features of the period between the end of the LGM and the start of the Holocene are the dramatic changes in climatic conditions after Heinrich event 1 and shifts in the climate variability, notably the sudden decline after the Younger Dryas. The scale of the warming in the Bølling/Allerød period provided considerable opportunities for the most adaptable communities, notably in southwest Asia in what is known as the Fertile Crescent, to exploit the abundance that came with warmer temperatures and greater rainfall. There is also a suggestion in the figures that climatic variability declined somewhat during these warmer spells (see Fig. 2.9b), which may have permitted the establishment of a more settled existence in favourable locations. Although the Older Dryas interrupted this climatic amelioration, the period of amelioration probably provided the opportunities for populations in favoured areas to rise. The savage cold and aridity of the Younger Dryas brought this period of advance to a grinding halt. The return to near-ice-age conditions posed the most frightful challenge to the burgeoning communities of the Fertile Crescent. Driven to oases and riversides, they may have found agriculture the only option for survival.

Recent archaeological studies have provided evidence of climate change at the end of the ice age that supports the Malthusian interpretation of the origins of agriculture. This new information has come from the site at Abu Hureyra, on the Euphrates, in what is now northern Syria.

This site, which was probably an example of the Natufian culture (see Section 5.8), was established around 13.5 kya. Here there is an unbroken sequence of archaeological evidence stretching from hunter-gatherer times to full-blown farming. Recent results extend the evidence of domestic cereals in the region, which involve forms of wheat and barley, back before the conventional date of around 11–12 kya (Hillman *et al.*, 2001). Now, it appears that systematic cultivation of cereals started at least as early as 13 kya: close to the beginning of the Younger Dryas.

The evidence indicates that hunter-gatherers at Abu Hureyra first started cultivating crops in response to a steep decline in wild plants that had served as staple foods for at least the preceding four centuries. The decline in these wild staples is attributable to a sudden onset of a drier, colder, more variable climate. Work by Gordon Hillman, of University College London, and his colleagues found that the wild seed varieties gathered as food gradually vanished, before the cultivated varieties appeared. Those wild seeds most dependent on water were the first to die out, then one by one by the hardier ones followed. So the hunter-gatherers turned to cultivating some of the foods they had previously collected from the wild. In an unstable environment, the first farmers started simply by transferring wild plants to more suitable habitats and cultivating them there.

The decreased annual yields of wild cereal stands drove people to adopt cultivation. This response probably occurred most naturally in areas where precipitation levels differed appreciably over short distances, notably in northern Mesopotamia. The other great advantage of this region was that several native grasses, notably wild barley, and einkorn and emmer wheat, grew in the mountains of eastern Anatolia and the Zagros Mountains. The process of domestication of these cereals was not the result of deliberate action. Higher yields came automatically from repeated cycles of sowing, growing and harvesting, which favoured heavier seeds and denser seed heads together with avoiding the propensity to shed seeds over an extended period. More deliberate action might have been a result of selecting grains that had less tough husks around the kernel, which would have been attractive as an aid to digestion.

It was not just a matter of grain cultivation. These radical effects of the Younger Dryas on human lifestyles are found in other evidence from the Levant. Hunting and gathering bands were forced to adapt to rapid climate change in the niches where wild plants and animals had formerly provided abundant subsistence. In the face of rapidly shifting conditions, subsistence strategies were extended to adjacent areas less affected by Younger Dryas drying and cooling. Archaeological findings reveal that the wild ancestors of all of the earliest, primary Neolithic crops grew together only in this Fertile Crescent region. Apart from the two types of wheat and barley, other plants that could be domesticated included lentils, peas and chickpeas. The wild chickpea provides a convincing key to the researchers' conclusions, as it is an extremely rare species occurring only in southeastern Turkey and northern Syria: had agriculture originated elsewhere, chickpeas would not have been domesticated as one of the first, or founder, crops.

While the earliest evidence of agriculture comes from the Fertile Crescent region, it is striking that the move towards agriculture occurred independently in other parts of the world at the beginning of the Holocene or even earlier. A consistent picture is emerging in South America, Mesoamerica, North America and China. Although cultivation may have been born first in the Near East, the latest evidence suggests that people on other continents began to domesticate the plants they lived with by the early Holocene – squash on the tropical coast of Ecuador (Smith, 1997) and rice along the marshy banks of the Yangtze in China (Shelach, 2000). Indeed the exploitation of rice may have pre-dated the Younger Dryas in East China. Fossil rice phytoliths have been identified from late glacial to Holocene sediments in the East China Sea that were probably transported by the Yangtze River from its middle or lower reaches (Lu *et al.*, 2002). The phytoliths appeared first in the sequence at about 13.9 kya and disappeared during the period of 13–10 kya, which includes the Younger Dryas. The disappearance during the colder conditions suggests that this frigid period had a significant influence on human activities at the time. Warmer and wetter conditions before and after this near-glacial interval probably favoured rice domestication in the area. These results have changed archaeological thinking about the dawn of agriculture, moving it several thousand years earlier in these regions. All this suggests that the dramatic changes of the Younger Dryas were driving this movement. After the improvements of the Bølling and the Allerød, the challenges of radical switches in the climate before and after the Younger Dryas, combined with the consequent disruption of animal migration patterns, formed the trigger that made agriculture the only means of survival for many people.

Later developments included domesticating millet around 6 kya in northern China, while rice appears to have been domesticated independently in southeastern Asia around the same time. In Mesoamerica, maize (*Zea mays*) began to be cultivated about 7 kya, and was the 'miracle' crop that enabled large civilisations to develop, while in Peru the bean was domesticated around the same time. So, while the shifting pressures of the climatic changes around the Younger Dryas may have provided the initial impetus for the adoption of agriculture, in the long run the less variable climate of the Holocene was the vital factor in its survival. As noted in Section 2.8, humans emerged from the climatic 'long grass' and confronted the relative tranquillity of the Holocene landscape, at which point agriculture was the only option.

5.8 NATUFIAN CULTURE

Another piece in the jigsaw of agricultural origins is the lifestyle identified as the 'Natufian culture' that thrived in the Levant between 14.5 to 12 kya (Bar-Yosef, 1998). During the first half of this period temperatures rose and precipitation increased and in the southern Levant reached a peak around 13.5 kya. This led to increased vegetation and greater yields of wild fruit, seeds and game animals, which altered lifestyles. In favoured areas sedentism became the preferred settlement pattern. With sedentism came population growth.

Archaeological studies of the Natufians provide extensive information on the transformation from a hunter-gatherer lifestyle to one of plant cultivation and animal domestication. Dorothy Garrod coined the term Natufian. She was a prehistorian at Cambridge University, who was the first woman to become a professor at the University in any discipline, and was responsible for a large number of excavations in Palestine from 1928 to 1934. Her findings came, in particular from a cave (Mugharet El-Wad) on the western side of Mount Carmel, near the town of Athlit, in present-day Israel. The excavations revealed more than one hundred individual human burials on the terrace directly in front of the cave. Some were found with ornamentation of bone, stone or shells.

Many of the flint tools found at the site were of a lunate from (i.e. shaped like the crescent of the moon). These could be used for a variety of purposes, but of particular interest were the larger ones that could be used as sickle blades attached to a wooden or bone handle, to form a scythe that could be used for harvesting cereals. Many of the tools show patterns of wear that confirm intensive and lengthy use for this purpose. In addition, there was a wide range of grinding equipment, including querns, pounders and pestles and mortars, which suggested great reliance on cereals and nuts for food. Other tools found include scrapers for treating animal skins, points for wood and bone working, awls for piercing, fish hooks and stones used as fishing weights, skins and decorative beads. There is also evidence of bows and corded fibres. An ornately carved deer scapula was found to have wear markings that indicated that it was used in the smoothing and straightening of wooden shafts, presumably for arrows.

Having thrived during the climatic amelioration of the Bølling and the Allerød, the Natufian culture then adapted to the Younger Dryas. This involved a shift from the largely sedentary existence to increased mobility, shorter-term occupation of sites and smaller social groups. The changes in populations have been inferred from archaeological analysis small game assemblages at various Natufian sites (Stiner & Munro, 2002). This shows that in the early period there was a much higher proportion of fast small game, whereas in the later period there was a greater reliance on slower prey (tortoises). Clearly, the population declined appreciably during the Younger Dryas.

The closing stages of the Younger Dryas marked the disappearance of the Natufian culture. It may seem surprising that these people did not flourish with the return of warmer wetter conditions, which coincided with the upsurge of agriculture in the Levant. This reaction underestimates the impact of a thousand years of colder drier climatic conditions. After developing a sedentary existence, the return to a nomadic lifestyle could have fragmented their culture. Nevertheless, the speed with which agriculture emerged with the climatic amelioration suggests that they and other groups had retained knowledge of the managed exploitation of cereals and other plants throughout the cold millennial drought of the Younger Dryas.

5.9 ÇATALHÖYÜK

It is a standard part of the presentation of the emergence of agriculture in the Fertile Crescent and the development of settlements to move on to Jericho as being the 'oldest city'. This is a logical way to proceed, and there is no particular desire here to usurp this primacy, but in terms of the nature of human social evolution, Çatalhöyük in central Turkey may represent a more interesting departure (Balter, 2001).[1] This was one of the earliest large farming settlements. It is an open question as to whether it ever constituted a 'town', let alone a 'city', in that it appears to have been a place where farmers lived, rather than somewhere that provided specialist skills and trades surrounded by occupied farmland. As such it seems to have been more of an overgrown village.

The striking feature of the settlement was its longevity. The latest radiocarbon dating estimates suggest that it was occupied from around 9.3 to 8.2 kya. It could conceivably have consisted of up to 5000 people in its heyday. Individual buildings were typically occupied for between 50 and 80 years, although it appears that they were actually occupied for a longer duration earlier on and that this decreased through time. Although this was a farming community its principal

[1] Details of progress on the Çatalhöyük excavations can be found on the website http://catal.arch.cam.ac.uk/catal/catal.html.

raison d'etre appears to have been that it was the centre for trade in obsidian that was mined locally. This trade extended as far as southern Palestine. As such it appears to have been a prototype for the trading centres that would come to form the earliest cities of the Middle East. At the same time there is little evidence of the handling of obsidian being concentrated in one part of the settlement. The widespread presence of knapping debris suggests that a significant part of the community was involved with the production of artefacts.

The timing of this occupation is interesting both in terms of the earliness of its original occupation and more so its abandonment. The emergence of the settlement appears to be consistent with the spread of agriculture within the Fertile Crescent. The latest estimate of its decline coincides with the sudden cold snap around 8.2 kya that brought much lower temperatures to the North Atlantic and Europe. Çatalhöyük was sited in a marshy area, which is slightly odd, as analysis of the phytoliths in the cereals consumed at the settlement indicates that the abundant wheat and barley were not grown in a wet alluvial landscape, but in drier, well-drained soils, the nearest of which were at least 12 km away. Similarly, the sheep and goats, whose bones are ubiquitous at the site, could not have grazed close to the town during the wettest months. Presumably the site was chosen to maintain adequate water supplies during the hot dry summer, and the fact that the principal sources of grain were from sites some 12 km from the town was an acceptable price to pay. The deterioration of the climate around 8.2 kya, which appears to have brought markedly cooler wetter winters, may have made the site less hospitable and led to its decline (Wick, Lemcke & Sturm, 2003).

The design of the buildings is another odd feature of the settlement. There were no streets. All the structures were connected, like a layer from a huge polygonal beehive. Entry to each dwelling was through an aperture in the flat roof, with access across the roofs of the adjacent buildings. The apparent benefit of this layout was that it was particularly easy to defend, or possibly more correctly difficult to attack. If so, it represents one of the earliest examples of a defended settlement, although there is no evidence of its ever having suffered a major assault.

Another interesting transitional feature of Çatalhöyük was the evidence of religion found in the dwellings. Clay figurines of what has been defined as both the 'Goddess of the seasons' and 'Mistress of the animals' hark back to the Venus figurines of Gravettian culture, although we must be careful about using such loaded terms to describe these figurines (see Section 4.7). At the same time a shrine adorned with bulls' heads seems to presage the more organised religions that were to emerge in subsequent millennia.

An equally profound consequence of these cultural developments is found in recent work that suggests that Anatolia, around 9 kya, is the cradle of Indo-European languages (Searls, 2003; Gray & Atkinson, 2003). Although we have eschewed any detailed discussion of the results of 'glottochronology' in discussing human evolution (see Section 1.2), this analysis does raise intriguing questions about whether the culture that emerged at Çatalhöyük defined our linguistic heritage.

5.10 PEOPLE AND FORESTS MOVE BACK INTO NORTHERN EUROPE

While the Middle East was experiencing changes that led to sedentism and the emergence of agriculture, Europe was undergoing more fundamental environmental shifts. The loosening of the ice-age shackles was, for the hunter-gatherers of southern Europe, a case of the Chinese curse: 'May you live in interesting times.' The ebb and flow of the forests during the Bølling, Older Dryas, Allerød and Younger Dryas posed major challenges for both animals and the humans who lived off them. The two major refugia were in southwestern France/Cantabria and Ukraine/Central Russian Plain, but other minor refugia could have existed in between. The way in which the western European refugia developed after the LGM and fuelled different expansion routes towards the north is, however, unclear. The timing of and interaction with the recolonisation coming from the eastern refugia also require examination.

The Magdalenians, who lived in southern France and northern Spain, had developed a remarkably sophisticated culture. They were superb artists, as the cave paintings of Lascaux and Altimira confirm. They were skilled workers in bone and ivory, producing barbed spears and harpoons, and had developed the spear thrower (*atlatl*). They spread northwards into Switzerland, southern Germany, Belgium and even the British Isles during the Bølling as willow and birch colonised the tundra and their traditional prey of reindeer, bison, horse and mammoth moved farther north.

The real challenge came after the setback of the Younger Dryas when the forests, having been driven southwards by the cold, started their ultimate spread northwards. The basic problem for people that relied on big game for sustenance was that forests provided much less food than the great plains with their vast herds of large mammals. As a broad indication of the difference, it is reckoned that forests can only provide around a quarter of the mammal biomass per unit area that open grasslands can. Living on the southerly edge of the massive Eurasian plain the hunters of postglacial Europe had access to a well-stocked larder. It is generally agreed that the extraordinary concentration of cave art in Cantabria and southern France is related to the high population density of humans in the region. Combined with the migratory routes, notably of reindeer, through the mountains of the region, it was an ideal place to exploit the ecological bonanza.

Confronted with challenge of the encroaching forests the inhabitants were faced with a stark choice: either they moved northwards with their vanishing prey or they stayed and adapted to the new conditions. The scale of the change should not be underestimated. It is a measure of how the population declined that the number of occupied sites fell to about a third of the level before the forests arrived. It is not clear whether this drop was simply a consequence of migration away from the area, or whether it reflects a more fundamental reduction in the number of people who could survive in the rapidly changing circumstances. What is clear is that the people who stayed changed their diet radically, shifting from one dominated by reindeer to a forest-oriented collation of red deer, wild boar and wild oxen.

There are two particularly interesting aspects of these migrations (Torroni *et al.*, 1998). Both relate to the extent to which the genetic evidence connects the Basques with other groups in modern Europe. The first is the link between the Basques and the Celts: what are sometimes referred to as the 'Atlantic populations'. The Y-chromosome complements of Basque- and Celtic-speaking populations are strikingly similar. In each of the Basque, Welsh and Irish populations, about 90 per cent of their chromosomes contain the M173-defined haplogroup (see Section 4.3; Wilson *et al.*, 2001). By way of comparison, a Turkish sample is much more diverse at the haplogroup level. There is no evidence, therefore, that the genetic similarities between the Basques and the Celts can be explained in terms of the contribution made by the people who moved into Europe during the Neolithic (see next section) or to later immigrants originating in the Near East. So it looks as if the Atlantic populations represent the least diluted form of the palaeolithic people who survived the LGM in southern France and northern Iberia.

The connection between the Basques, the pre-Anglo-Saxon British and the Irish has interesting implications for the northward expansion from a glacial refugium in northern Iberia. It is consistent with the observed diffusion of Magdalenian industries in western Europe. In practical terms, people living in Cantabria at the end of the LGM, when the climate started to warm up rapidly, could have walked to Ireland over the continental shelf exposed to the west of what is now Brittany and then across the English Channel and the Celtic Sea. Even after the Younger Dryas, the land bridges existed, although there may well have been a narrow strip of water between Ireland and the rest of the continent. Then between 10 and 8 kya the British Isles would have been formed and cut off from the continent of Europe, thereby largely establishing the genetic isolation of the Celtic people that was not really challenged until the arrival of the first Romans and then, more importantly, the Anglo-Saxons. At the same time, the thick forests that expanded across northern Europe acted as a barrier to the large-scale migration of people for much of this time. While the concepts of agriculture and other technologies spread across the continent, the spread of people was less substantial.

The other, rather more surprising connection is between the Basques and the Saami peoples of northern Finland. The genetic evidence shows that a population living in the Iberian peninsula and southern France before the Younger Dryas has contributed substantially to the gene pool of all modern populations of central-northern Europe, including the Saami and the Finns. This late Palaeolithic population expansion from southwestern to northeastern Europe is supported not only by mtDNA data and Y-chromosome data, but also by archaeological records.

The reason that this is unexpected is that the Saami, together with a few other European populations including the Finns, speak a language belonging to the Uralic linguistic group. This suggests that these populations originated in western Siberia and admixed with European populations. An almost completely European origin of the Saami and Finnish mtDNA and Y-chromosome markers contradict this hypothesis. They show genetic patterns that belong to European haplogroups. In contrast with the Saami and Finns, a large majority of the markers of Siberian populations are of Asian origin (Torroni *et al.*, 1998; Tambets *et al.*, 2004).

This suggests that the majority of the ancestors of the Saami were effectively the most dedicated of the reindeer hunters living in the southwestern France/Cantabria refugium. They continued to pursue their prey over the millennia, as they moved ever northwards. As part of this evolving relationship the simple hunting of the herds as they moved through a given area developed into the semi-domesticated lifestyle for which the Saami are now so famous.

Another intriguing aspect of the changes at the beginning of the Holocene is the almost complete cessation of artistic output, be it cave paintings or mobiliary art. This suggests that, as part of the climatic upheavals at the time, the decline in population density and possibly living standards altered the perceptions of either the need for or capacity to produce these artefacts. The parallel decline in the quality of the antler, bone, ivory and stone tools also implies a lack of time or inclination to concentrate on the artistic nature of the artefact rather than its purely functional requirements.

The change that occurred when the forests returned to most of Europe at end of the Younger Dryas set the seal on the future of human activities across the continent. Pollen diagrams, which show the spatial and temporal spread of trees across Europe, paint a picture of an invasion that lasted a few thousand years. In the vanguard were birch and pine which reached Denmark by 10.5 kya, and were closely followed by hazel and then elm. Then around 8.5 kya came lime, oak and alder. The pattern of advance varied as oak already covered the southern half of the continent by 10 kya and reached its northerly limit in Britain by 8 kya and in Scandinavia and Russia by 7 kya (Fig. 5.1). Lime took a somewhat different route. Starting from a smaller area in the Balkans and Italy, it reached its northern limit by 7 kya (Fig. 5.2). Overall, the transition to peak Holocene levels of tree cover in the eastern Mediterranean took about a thousand years following the end of the Younger Dryas. Farther north the forest cover was still rather more open than at present with more herbaceous glades, but by 8.5 kya the forest had become closed.

Genetic mapping provides new insights into faunal developments. The colonisation of northern Europe by smaller creatures followed a variety of routes, depending on the scale of the barriers presented to them by the Pyrenees and the Alps. In the case of the two species of European hedgehog (*Erinaceus europaeus* and *concolor*), there were parallel migrations northwards from refugia in northern Spain, Italy and the Balkans, to occupy three broad north–south zones. In contrast, the colonization of northern Europe by the grasshopper (*Chorthippus parallelus*) sprang exclusively from the Balkans (Hewitt, 2000), as the mountains farther west formed too great a barrier to their movement.

There is an interesting question about the impact of the 8.2 kya event on European developments. It is not evident in the pollen

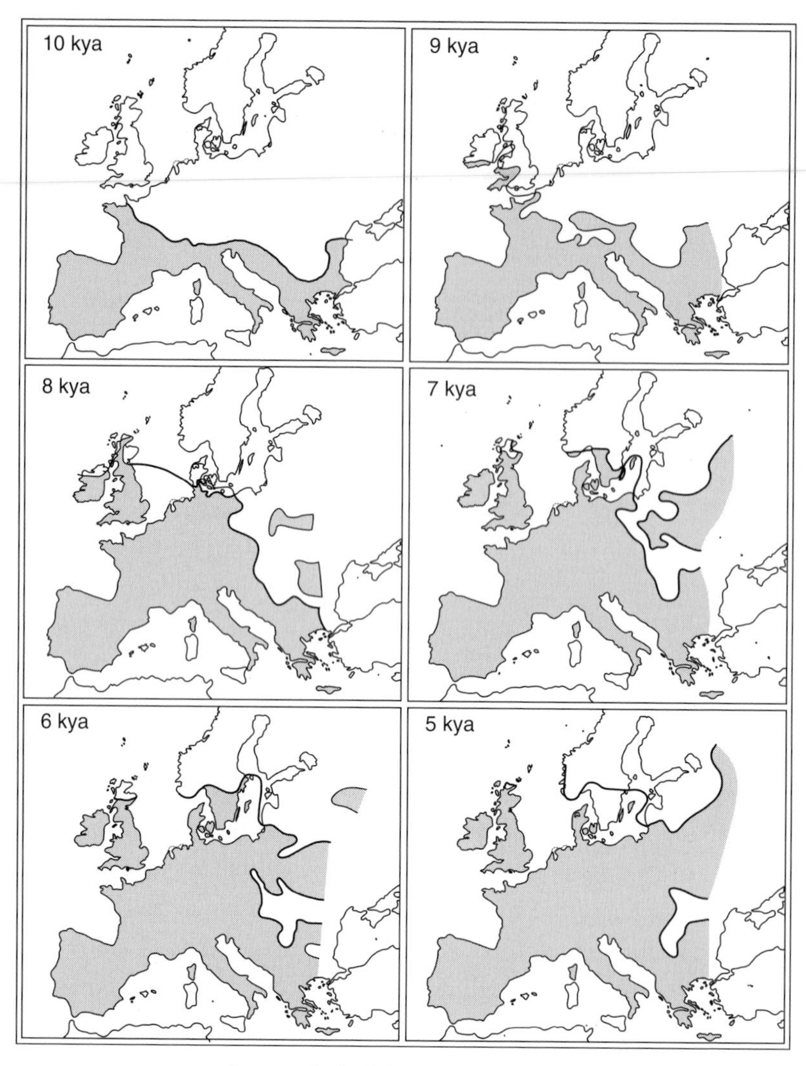

FIGURE 5.1. The spread of oak (*Quercus*) across Europe during the early to mid-Holocene.

diagrams, but can be seen clearly as a sharp cooling in the Ammersee record (see Fig. 2.8, and Section 5.2). At about the same time, there appears to have been a marked decline in the extent and number of Mesolithic settlements in southern Germany (Mithen, 1994, p. 119). This decline is also identified as a conspicuous gap in the radiocarbon

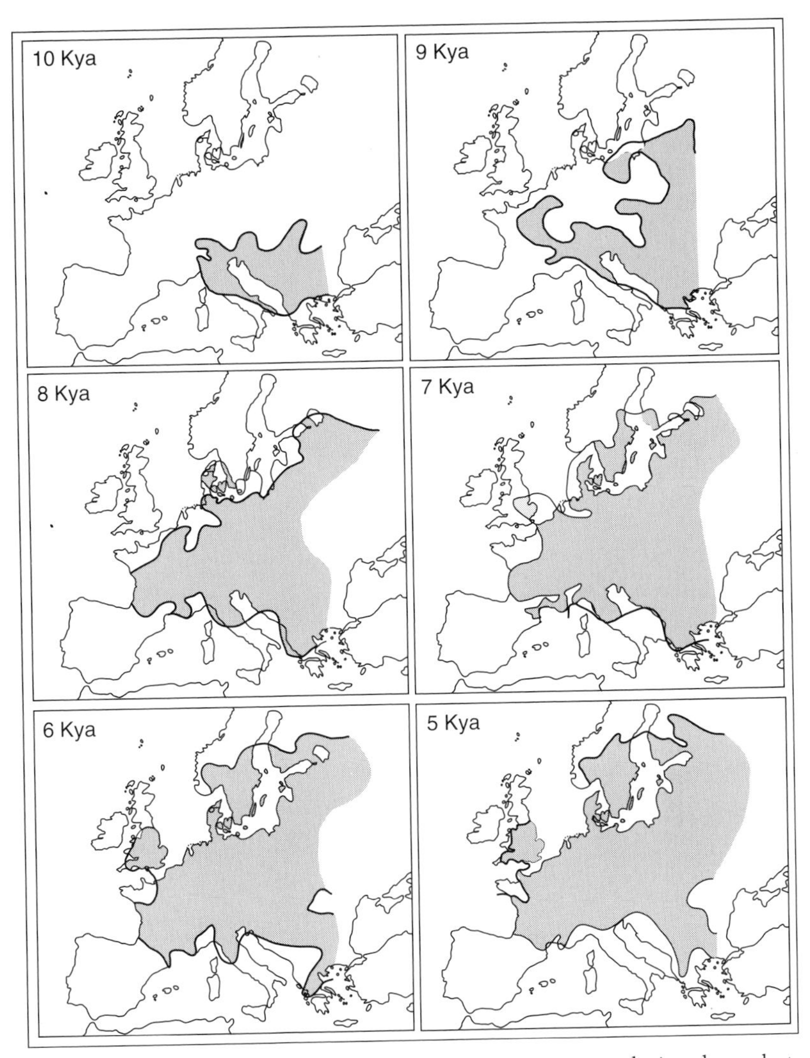

FIGURE 5.2. The spread of lime (*Tilia*) across Europe during the early to mid-Holocene.

record of settlements around the Iron Gates on the River Danube (Bonsall *et al.*, 2002). Many riverbank sites were abandoned between about 8.25 and 7.9 kya. This abandonment is linked to increased flooding along the Danube, which fits neatly with the cooler wetter climate of the 8.2 kya event. Combined with the abandonment of both sites in southern Germany and Çatalhöyük, this all suggests that climate-related flooding had a significant impact on human use of riverine environments in southeast Europe at this time. In addition, the conclusion of the Danube analysis is that the cool event delayed appreciably the spread of agriculture in southeast Europe.

The pollen records of the mid-Holocene contain dramatic evidence of the environmental change between 6 and 5 kya. They show that in Europe around 5.7 kya there was a severe decline in elm (*Ulmus*), while in North America the hemlock (*Tsuga*) went into decline around 5.3 kya. These declines have been attributed to specific pathogen attacks, but it is possible that climate deterioration played a part, if only to increase the stress on certain species and ease the spread of epidemics. The alternative explanation is that the Europe-wide 'elm decline' was the result of a sudden increase in agriculture and herding, with Neolithic farmers cutting down elms to feed their livestock. This seems unlikely. Even in Greece, significant soil erosion due to deforestation does not seem to have occurred until after 4 kya, so it is hard to see how deforestation could have had an earlier impact on the vast forests of northern Europe.

5.11 THE SPREAD OF FARMING INTO EUROPE

The next profound change in northern Europe following the spread of the forests and the colonisation by Mesolithic peoples was the arrival of agriculture. Here is yet another area where the academic world is locked in vigorous debate as to the exact nature of the change. The archaeological evidence provides a broad picture of farming spreading out of the Middle East during the early Holocene. By 9 kya it had reached Greece, and by 6 to 5 kya it was up into Britain and Scandinavia (Fig. 5.3). The forests had already moved north across Europe and new cultures had become established that depended on exploiting a wide range of flora and fauna. The bone of contention is whether, in the orderly progression from hunter-gatheres society to agriculture, the people who made this revolutionary transition were the descendants of the hunter-gatherers who had survived the ice age and the upheavals that followed it, or of invaders who dispossessed them. In short, did

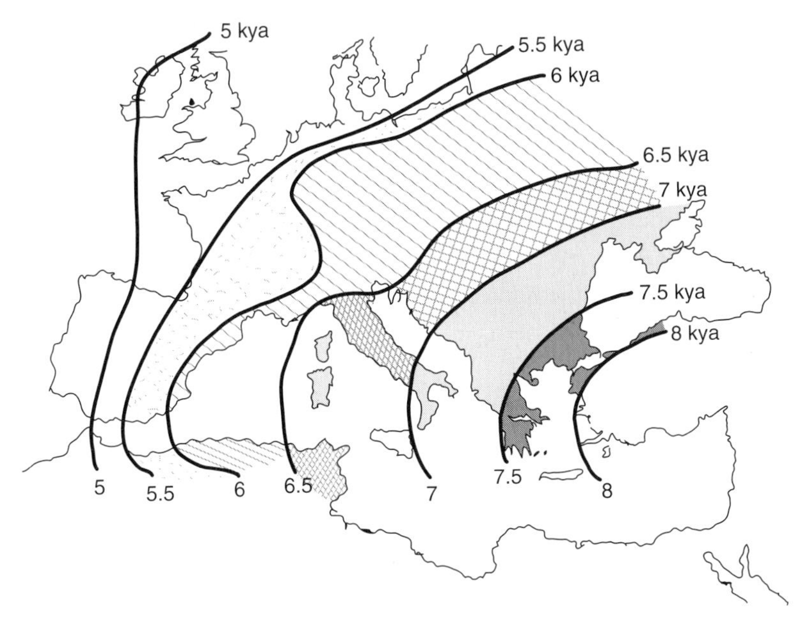

FIGURE 5.3. The spread of farming across Europe during the early to mid-Holocene.

farmers move into Europe from the Fertile Crescent, or did the locals survive and learn to trade and farm for themselves?

The earliest glimpse of European genetic origins came from protein markers (Cavalli-Sforza, Menozzi & Piazza, 1993). This work was followed by research into the mtDNA of European women. Although these studies showed that several European genetic markers were too old to have been introduced with the Neolithic newcomers, there was a heated debate about the extent to which this was the case. Because the genetic maps of Europe were so similar to the model of the spread of farming from the Middle East, it was argued that most of the change had to be associated with the large-scale influx of immigrants, who came to constitute the principal source of modern Europeans.

Genetic mapping studies of both mtDNA and the Y chromosome of living Europeans have provided the same answer to this fundamental question (Semino *et al.*, 2000; Underhill *et al.*, 2000; Ke *et al.*, 2001; Wells *et al.*, 2001; Sykes, 2002). More than 80% of

European men have inherited their Y chromosomes from Palaeolithic ancestors who lived 25 to 40 kya. Only 20% of Europeans trace their Y-chromosome ancestry to Neolithic farmers who moved in from the Middle East. The data from both genetic lineages not only enabled researchers to trace the movements of the first farmers, they also paint a detailed picture of the identity and prehistoric movements of ancient Europeans (see Section 4.3).

The final migration has been identified by four new mutations and occurred about 9 kya. Only about 20% of Europeans have these Neolithic markers. The distribution of markers even suggests something about the route the ancient farmers took. There are more Palaeolithic markers in the north of Europe than the south and more Neolithic in the south. This may be a sign that at least part of the Neolithic people went by boat along the coast of the Mediterranean.

This set of apparently clear lineages has been thrown into doubt by more recent work by Lounes Chikhi, from University College London. Chikhi and colleagues looked at different markers on Y chromosomes (Chikhi *et al.*, 2001). In particular, they studied rare mutations called unique event polymorphisms (UEPs). These are not thought to have occurred more than once in recent human history. The presence of UEPs in different populations is likely to indicate common ancestry rather than recurrent changes in gene structure. The research team took the results of a previous study and subjected them to a new computer-intensive technique. From this, the scientists estimated that Middle Eastern farmers contributed about 50% of the analysed genes to the modern European population.

Contributions ranged from as little as 10% in western France and England, to 85–100% in southeastern European countries such as Albania, Macedonia and Greece. These percentages are much higher than previous estimates, suggesting that the Middle Eastern contribution to European genetic heritage has been underestimated. Particularly interesting is the apparent migration around the coastal areas of the Mediterranean, where the genetic evidence of the influx from the Near East appears to be stronger. The results suggest that the spread of agriculture in Europe was due to an influx of Middle Eastern farmers, which supports the Demic Diffusion Model (see Section 4.3). Clearly, this debate about the lineage of Europeans has some distance to run, although at present the balance of the evidence is that about three-quarters of the present-day European genes are descended from indigenous Palaeolithic ancestors.

Whatever the outcome of the overall debate about the peopling of Europe, the available evidence supports the hypothesis concerning the connection between the Basques and the Celts (see Section 5.10). The closeness of the Basque and Irish Y-chromosome data supports the case for the relatively small Neolithic component in these populations. These data support the proposition of a close relationship in the paternal heritage of the Basque- and the Celtic-speaking populations of the British Isles. Furthermore, this suggests that here the Neolithic transition did not entail a major demographic shift and farming spread more through cultural transmission than through migration.

5.12 THE PEOPLING OF THE NEW WORLD

The debate about the genetic origin of people in Europe is as nothing compared to that concerning the timing and means by which the New World was initially peopled. This is one of the most fiercely contested areas of archaeological research. In the late 1920s it was conclusively demonstrated that humans coexisted with late Pleistocene megafauna when a number of distinctive bifacial flint spear points were found embedded in the bones of an extinct form of bison at Folsom, New Mexico. Subsequently, an earlier form of these artefacts was found at Clovis, New Mexico, and they became known as Clovis points. These discoveries established an orthodox view that no unequivocal evidence for the peopling of the New World existed before the Clovis horizon. This horizon was measured as being between 13.2 and 12.9 kya. Given this seemingly late date for the arrival of the so-called 'First Americans', conventional wisdom has also maintained that the initial migration through Beringia to the Americas could not possibly have occurred before about 14 kya.

So entrenched did the 'Clovis First' hypothesis become that those who had doubts about its ascendancy claimed that it was tantamount to academic suicide to publish work that contradicted this orthodoxy. In exploring this debate and considering the climatic evidence it is important to restate the approach adopted throughout this book. Here, the objective is to present the evidence of climate change and the climatology in prehistory that may assist in addressing debates of this type, not to seek to provide answers to the debate. In some cases, however, this does mean asking whether the climatic arguments have been adequately integrated into the analysis, and whether this may steer us in one direction or the other.

In the case of the peopling of the Americas, climatic arguments dominate the classic analysis. In the case of the peopling of the Americas, climatic arguments dominate the classic analysis. Put at its simplest level, the only feasible route into the continent was assumed to be from eastern Siberia via northern Canada. This was closed during the LGM. Since there appeared to be no evidence of a human presence from well before 13.5 kya, it was argued that humans could not have arrived 10 to 20 kyr sooner, when the corridor between the Cordilleran and Laurentide ice sheets was open (see Section 3.3). Nevertheless, the general assumption is that the extent of the Cordilleran and Laurentide ice sheets prior to around 30 kya (see Section 3.3) was such that there could well have been a substantial corridor into North America via northern Canada. Furthermore, the recent evidence of human occupation of northern Siberia as early as 40 kya and within 2000 km of the Bering Straits by 30 kya (Pavlov, Svendson & Indrelid, 2001; Pitulko *et al.*, 2004) means that people could have reached Beringia by 30 to 25 kya. This means that recent claims of humans arriving sometime before or even during the LGM cannot be dismissed on climatic grounds alone.

One of the reasons for the reluctance to accept pre-Clovis dates is that the Clovis phenomenon appears to have been a continent-wide, west-to-east colonisation by highly mobile, specialised big-game hunters. The Clovis hypothesis used emotive expressions like *overkill* or

Blitzkrieg to describe the arrival of human populations at 13.5 kya and their rapid spread throughout the entire hemisphere. Such a model made the verification of putative pre-Clovis localities difficult because the hunters had not left an adequate trail of their occupation. In particular, the dramatic changes that occurred in environmental conditions across North America during the emergence from the ice age (see Section 5.5) would have required a variety of foraging strategies and also erased much of the evidence of these lifestyles. In questioning the dating of any pre-Clovis sites the debate has been acrimonious and criticism of the results virulent.

Efforts to convince a sceptical archaeological establishment have drawn on three principal sources: new archaeological material of great antiquity, genetic profiling of Native Americans and links between the distinctive Clovis points and similar artefacts in the Old World. A good example of the first type of evidence is the site of Monte Verde in south-central Chile. On the basis of carbon dating of the exceptionally well-preserved organic materials and artefacts, the occupation of the site been identified as being between 14 and 14.5 kya.[2] This dating can only be regarded as pre-Clovis. These results produced a storm of argument that centred on whether the artefacts were real and the dating accurate. The investigators, under Tom Dillehay from the University of Kentucky, produced a magnum opus that appears to have adequately met the objections to the dating of their work (Dillehay, 1997; Grayson, 1998; Dillehay, 2000). It is, however, in the nature of this particular debate that it continues to rumble on generating more heat than light.

A further refinement to this debate was a paper in 1986 combining anthropological, archaeological and linguistic studies. This concluded that there had been three separate early migrations from Asia

[2] In some early texts there is confusion about the age of this site as dates are quoted in radiocarbon years. At the time in question, this made about 2000 years' difference. So, much of the discussion of how Monte Verde fitted into the 'Clovis first' debate used a date of 12 to 12.5 kya (the radiocarbon dates), which could just about be squeezed into the post-ice-age chronology, whereas the corrected calendar dates of 14 to 14.5 kya cannot be reconciled with Native Americans entering the New World from Beringia by a land route.

to the New World. Called the 'Greenberg hypothesis' after Stanford University linguist Joseph Greenberg (Greenberg, Turner & Zegura, 1986; Greenberg & Merritt, 1992), it proposed that three waves of migrants, starting around 12 kya, established three different genetic patterns that corresponded to three different language families. These three groups were the Eskimo-Aleut, who settled in Alaska, northern Canada and Greenland; the Na-Dene, who now live in parts of Alaska, northwest and north-central Canada, and in the American southwest and northern Mexico; and the Amerind who populate the rest of the New World. This hypothesis has been the subject of intense debate since its publication.

The contribution of genetic profiling to the debate has had mixed results. Studies of mtDNA of Native Americans, published in 1992, by Douglas Wallace and his colleagues at Emory University in Atlanta, found four distinct lineages (Wallace & Torroni, 1992). These were designated A, B, C and D (see Fig. 3.6). These lineages were also found in Asian populations, but not in Europeans and Africans. Furthermore, all four lineages appear in Amerind people, but in Na-Dene only A is present, while in Eskimo-Aleut people only A and D appear. These genetic studies concluded that there was a major wave of northeastern Siberians who entered the Americas between 25 and 20 kya. This corresponded to the Amerind group. About 12 kya another wave of people moved into the Americas. This second wave was the ancestors of the present Eskimo and Na-Dene populations.

The new genetic model has had to compete against alternative interpretations of both its genetic evidence and the collection of much more data from across the Americas. Analysis of the level of mutation in both the mtDNA and Y chromosomes of Native Americans provided a mean date of their separation from their Asian forebears as being in the range 17 to 34 kya. As for the number of migrations, there is no consensus. The studies of mtDNA have produced a variety of competing scenarios ranging from one to six separate waves (Torroni *et al.*, 1993; Starikovskaya *et al.*, 1998; Stone & Stoneking, 1998; Lell *et al.*, 2002). The Y-chromosome data have supported either the three-migration model or a four-migration pattern. Furthermore, there are different proposals for which 'source' populations in Asia gave rise to New World populations: Y-chromosome data implicate Mongolia/ Manchuria and/or extreme southeastern Siberia as the ancestral homeland; whereas mtDNA data point to Mongolia, North China, Tibet and/or Korea as the candidate source regions in Asia. More specifically, it has been argued that the first wave of immigrants most closely resembled the prehistoric Jomon people of Japan (see Section 3.11) and their closest modern descendants, the Ainu, from the island of Hokkaido. The Jomon and Ainu have skull and facial characteristics more physically similar to those of Europeans than to mainland Asians.

The possibility that the peopling could be the product of a single wave, albeit one that took place in dribs and drabs some 25 kya, is particularly interesting. It is the product of work by Andrew Merriwether, and colleagues, when at the University of Pittsburgh, on mtDNA samples from 1300 Native Americans representing more than 40 populations throughout the Americas, along with 300 samples from teeth, bone fragments and mummified tissue at three burial sites, one each in North, Central and South America (Merriwether, Rothhammer & Ferrell, 1995). By including a broader range of populations and large sample sizes, this research provides more detail of the evolutionary relationships among Native Americans. It shows that all four lineages – A, B, C and D – have been found in people from all three linguistic groups, although lineages B, C and D are rare in Na-Dene, as are B and C in Eskimo-Aleut.

The explanation of how a single, extended migration could have ended up looking like three waves depends on the people being split up into small isolated groups soon after their arrival. The group that headed farthest south became the Amerinds, while the others remained in the north close to the Laurentide ice sheet. As a result the Amerinds prospered and populated much of the continent. The others only just survived in isolated groups until the end of the ice age, after which they prospered in northerly latitudes. This bottleneck had, however, reduced their genetic diversity, and gave the impression that they had arrived later. Those who propose the multiple wave models, notably Douglas Wallace, did not, however, accept this hypothesis.

Climatic observations can assist in resolving this hotly disputed set of issues in two ways. The first relates to the timing of both the closing and opening of a corridor existing between the eastern flank of the Cordilleran ice sheet and the main Laurentide ice sheet.[3] This may have remained open well into the LGM, but it is exceedingly unlikely that it was used as a migration route at this time. To start with, it must have been an exceedingly inhospitable prospect for migrating bands of people, being almost devoid of vegetation and with few animals to hunt. In addition, there is no archaeological evidence of any human occupation in the vicinity of either end of the route at any time before the LGM. Where the climatic evidence is more intriguing is in respect of the opening of the corridor. The consensus view is that it did not open until around 14.5 kya. The archaeological evidence suggests that people started to colonise East Beringia by around 13.5 kya, but there is no evidence of colonisation of Alberta until 12.5 kya at the earliest: all of which makes it impossible for the same people to have reached southern Chile by 14 kya at the latest. So, unless the dates for Monte Verde are flawed, which only the diehard 'Clovis First' lobby insist is the case, some people must have entered North America by another route.

The alternative route for people moving into the Americas during the LGM is by boat. Although the North Pacific would have been appreciably colder than the present, because it does not exhibit the same strong thermohaline vertical circulation as the North Atlantic it did not undergo such dramatic changes in circulation. So, the pack ice down the coast of what is now Alaska and British Columbia did not represent too formidable a barrier to movement, given that the same humans that were to spread northwards through East Asia and

[3] The observations presented here are taken from the various papers that appeared in *Quaternary International*, **32**, pp. 3–128 (1996) on the Ice-free Corridor, and also from Mandryk *et al.* (2001).

eventually into North America had colonised Australia by 60 kya (see Section 3.8). Furthermore, the Cordilleran ice sheet did not build up into a substantial barrier until around 20 kya, and it then retreated to expose the coastline by around 15 kya (Mandryk *et al.*, 2001). As a consequence, the deeply indented coastline that was exposed by the lower sea level at the time would have provided a haven for groups moving along the coast. What is more, the shoreline and coastal waters would have provided abundant sources of food. Compared with the barren trek from the Canadian arctic through the narrow gap between the massive decaying ice sheets and around the lakes they formed, this coastal route seems like an altogether easier option.

This coastal route is supported by the interpretation of genetic analysis (Starikovskaya *et al.*, 1998), which suggests that lineages A, C and D of the people who moved into the Americans have a divergence data between 34 and 26 kya. These people may have occupied the coastal regions of Beringia as early as 34 kya. On the other hand, lineage B split off from the Siberian population between 16 and 13 kya. These may have been the mammoth hunters of central Beringia that then moved into North America bringing with them the Clovis technology. What is certain is that reconciliation of the archaeological and genetic evidence will take a lot more research.[4]

During the early and mid-Holocene the fluctuations in the climate described in Section 5.5 do not, for the most part, appear to have exerted a major influence on the peopling of North America. Changes in weather patterns during the early Holocene (see Section 5.5) may, however, explain two Amerindian demographic shifts in the northwest plains of North America. The first is that at around 11 kya, Amerindian culture apparently split into foothills-mountains and plains communities (Lovvorn, Frison & Tieszen, 2001; Foreman, Ogelsby & Webb, 2001). Then, from 8 to 5 kya, there is a scarcity of archaeological sites on the open plains, which had been attributed to an emigration because of

[4] An up-to-date and balanced assessment of the current state of the debate is provided by Meltzer (2004).

a period of aridity. The latest thinking is that the significant warming after the Younger Dryas led to the first cultural separation. The scarcity of archaeological sites after the 8.2-kya event is best explained by rapid climate fluctuations after the catastrophic draining of Lake Agassiz, which altered monsoonal inputs to the open plains.

5.13 CONCERNING BROWN BEARS AND HAIRLESS DOGS

Examining the genetic diversity of Native Americans is only part of the story. Geneticists have also done detective work on certain animals to draw conclusions about the timing of peopling of North America. These insights come from such unlikely sources as the sequencing of the genes of both brown bears and hairless dogs.

The study of the population genetics of modern and permafrost-preserved brown bears in western North America provides interesting insights into how both animals and humans survived the LGM in this part of the world (Leonard, Wayne & Cooper, 2000). This analysis shows that around 35 to 40 kya there were three genetic lines of bears in Alaska. There is, however, no evidence of brown bears living to the south of the ice sheets at this time. By 15 kya the population had declined to a single line in Alaska, which appears to have been on the verge of extinction. The current population in western Canada and northwestern United States belongs to a lineage, that was present in Alaska before the LGM, but is no longer in existence there. The only way that these changes in distribution can be explained is that some brown bears migrated down the west coast of North America between 35 and 15 kya to form the line for the southern enclave. The obvious inference is that what was possible for bears would have been an option for humans: both these omnivorous creatures took the same escape route from the worst of the ice age in Beringia.

The whole question of the antiquity of the relationship between dogs and humans has been the subject of debate within the genetic mapping community (see Sections 3.15 and 4.10). It may all seem a bit far-fetched, but the study of the wholly unpronounceable Xoloitzcuintli (Mexican Hairless Dog) provides an extraordinary insight into both human and canine history (Vila, Maldanado & Wayne, 1999). This medium-sized hairless dog was used by the pre-Columbian people for food, companionship and to relieve the pain associated with arthritis. It also had a religious status. When the Spanish conquistadors arrived in Mexico they brought their own dogs, and crossbreeding swamped the native breeds. They also systematically eliminated native dogs as part of a programme to replace local traditions with Hispanic culture. Because of its religious status, the people saved the Xoloitzcuintli by hiding them in mountain villages in western Mexico, where their breeding was carefully managed.

The history of the Xoloitzcuintli made a particularly useful candidate for exploring questions about the genetic origins of dogs in North America: did they come with the first people from Beringia, or were they domesticated North American wolves? Genetic analysis of Xoloitzcuintlis found there was no connection with New World wolves. The only common haplotype in this breed that is found in both wolves and dogs was identified as one presently found in a species of wolf in Romania and European Russia. The difference in the genetic profiles of the hairless dogs and the Romanian wolves suggests they separated some 40 kya. Given when the first Americans arrived in North America, the fact that dogs came with them suggests that the timing was closer to the split between wolf and dog rather than more recently, otherwise the Xoloitzcuintli would have closer links with a wider range of Old World wolves. This observation also supports the general case for active connections between dogs and men extending well back into the ice age. Furthermore, it reinforces the pre-Clovis implications of much of the genetic mapping of Native Americans.

5.14 A EUROPEAN CONNECTION?

An even more contentious aspect of peopling the Americas relates to tracing the origins of the technology that produced the Clovis points and their successors. Here, it is easy to get into much deeper water. The distinctive, beautifully crafted, bifacial fluted flint spearheads are clearly a feature of the people who moved into North America. There is, however, virtually no evidence of similar artefacts in Siberia. By way of contrast, an almost identical form of technology existed in southwestern France between 21 and 17 kya, associated with the people known as Solutreans (see Section 3.9). This coincidence has led a proposal that maybe these people, fleeing from the full rigours of the LGM, somehow crossed the Atlantic and became among the earliest settlers in North America (Stanford & Bradley 2000).

The possibility of people travelling down the west coast of the Americas can be invoked to explain how the Solutreans could have travelled across the Atlantic. Although this is an altogether greater challenge, the same skills of hunting sea mammals and fishing could have enabled bands of people to move westwards across the top of the North Atlantic using boats to sail round the edge of the pack ice. What is less clear is why these people should have embarked on this perilous and seemingly hopeless journey. Confronted by the increasing harshness of the LGM, which may have made much of southwest France uninhabitable, the obvious alternative was to move south. Since there is no reason to believe that the ice field on the Pyrenees shut off this escape route, it is hard to see why they took on the far greater challenge of crossing the North Atlantic. The fact that during the ice age the Gulf Stream would have been flowing roughly towards northern Portugal compounds the challenge of crossing the largely ice-covered ocean to the north of 40 to 45° N. The possibility of a southern route to the Caribbean does not appear to be regarded as a realistic alternative.

This seemingly far-fetched idea, which could be described as moving into the 'fringe archaeology' that was disavowed in Section 2.12, has, however, received a shot in the arm from two areas of research. First, mtDNA profiling by Douglas Wallace's group of Native Americans living in the Great Lakes region shows the existence of a fifth genetic lineage (Brown *et al.*, 1998). This form (X) only exists amongst Europeans and is not present in East Asians (see Fig. 3.6). The data suggest that this haplogroup arrived in the Americas either 12 to 17 kya or 23 to 36 kya. While the later date could be used in support of the Solutrean migration, the earlier date would rule out the Solutreans crossing the North Atlantic.

A possible explanation of the presence of the X haplogroup in North America is that it is a remnant of the people who reached northern Siberia between 40 and 30 kya (see Section 3.12). Then, driven forth by the rigours of the LGM, they migrated both east and west, while leaving no residual population in Siberia. The alternative is the somewhat unlikely proposition that a group of European women managed to migrate back all the way to North America via Beringia without leaving any traces of their genes in transit. Either way, we have some awkward questions to answer about how this genetic line is now found only in western Europe and the eastern United States. The second issue is the increasing evidence of Clovis-like artefacts being found at Cactus Hill, in the eastern United States, at levels that have been dated to as much as 16 kya. So there are still plenty of unresolved questions about the peopling of the New World.

5.15 FLOOD MYTHS

The change in sea level (see Section 2.10) since the end of the ice age has had a profound effect on both the geography of many countries and the lives of their people during the time since then. All around the world, large areas of continental shelf have been inundated in the past 15 kyr. Many of these areas offered attractive sites for human occupation, and it is possible that in some places surprisingly well-developed communities may have emerged, only to be flooded out by the advancing sea.

Although the rise in sea level may have been frustrating for archaeology, the important climatic message is that the inundation was an inexorable process lasting many thousands of years and, for the most part, not a cataclysmic event for humans. Individual communities may have been swept away by sudden storm surges driven on by strong winds, but the important underlying factor was the steady rise of sea level that made such final acts inevitable. This combination of wind, rain and rising sea levels remains a terrifying threat to communities from Bangladesh to Florida or around the North Sea. What is less clear is whether a succession of these events was sufficient to generate the flood myths that are so prevalent in many cultures, or whether some more dramatic events had a more indelible impact on people's thinking.

The extraordinary prevalence of the Noah Flood Myth suggests that something more than rising sea levels inspired such widespread records. The best known of these stories concerns Gilgamesh, who was king of Uruk, on the River Euphrates (see Section 6.3), and lived about 4.7 kya. Many stories and myths were written about Gilgamesh, some of which were recorded as early as 4 kya. These were integrated into a longer poem, versions of which survived in various languages written in the cuneiform script, and which has become known as *The Epic of Gilgamesh*. At one stage in the epic Gilgamesh, in seeking eternal life, undertakes a perilous journey to Utnapishtim and his wife, the only mortals on whom the gods had conferred eternal life. He was the great king of the world before the Flood and he and his wife were alone preserved by the gods during the Flood.

Utnapishtim tells how the gods resolved to destroy the world in a great flood. All the gods were under oath not to reveal this secret to any living thing, but Ea (one of the gods that created humanity) came to Utnapishtim's house and told the secret to the walls of the house, thus not violating his oath to the rest of the gods. He advised the walls to build a great boat, its length as great as its breadth, to cover the boat, and to bring all creatures into the boat. In his tale, Utnapishtim builds the great boat, which he then loads with gold, silver and all the creatures of the Earth. Ea orders him into the boat and commands him to close the door behind him. The black clouds arrive; the Earth splits like an earthenware pot, and all the light turns to darkness. The Flood lasts for seven days and seven nights. When, finally light returns, Utnapishtim opens a window and the entire Earth has been turned into a flat ocean; all humans have been turned to stone. His boat comes to rest on the top of Mount Nimush; the boat lodges firmly on the mountain peak just below the surface of the ocean and remains there for seven days. On the seventh day:

> *I [Utnapishtim] released a dove from the boat,*
> *It flew off, but circled around and returned,*
> *For it could find no perch.*
> *I then released a swallow from the boat,*
> *It flew off, but circled around and returned,*
> *For it could find no perch.*
> *I then released a raven from the boat,*
> *It flew off, and the waters had receded:*
> *It eats, it scratches the ground, but it does not circle around and return.*

Clearly, this myth is the origin of the story of the Flood in the Book of Genesis, which it predates by at least a millennium. But as has been made abundantly clear in Chapter 2, the available climatic records do not contain evidence of some global cataclysm happening between about 15 and 5 kya when most of the rise in the oceans occurred. There is nothing in the many climatic records to support various theories of sudden huge rises in global sea level associated with cosmic catastrophes or sudden shifts in the Earth's axis or its crust. Nonetheless, the widespread Flood Myth appears in many fables from prehistory.

There are two possible explanations of this story. The first is that it is simply a folk memory of the sea-level rise during prehistory. This is unlikely as, in human terms, the rate of rise was slow. Figures of around a metre a century would not have had shoreline communities rushing to the hills. The inexorable rise would, however, have combined with occasional storms to sweep away habitation close to the sea. This is sometimes sufficient to produce a folk memory of drowned communities. For instance, more recent changes around the British Isles have led to local stories of the ghostly bells of sunken villages still being heard on stormy nights. It cannot be claimed, however, that these have the flavour of Noah's Ark, or Gilgamesh.

The steadily rising sea levels might have had a more profound effect on coastal communities where large areas were inundated in fits and starts. For example, this could have happened in the Persian Gulf. This enclosed sea goes no deeper than 100 m, and much of the seabed is only about 40 m below the present-day surface. When sea levels were 120 m lower the gulf would have been dry land 20 kya, and the ancestral river system of the Tigris and Euphrates flowed through the deepest part of the gulf, a canyon cut by the river waters to the Indian Ocean. The postglacial rise in sea level inundated the floor of the gulf between 15 and 6 kya. The sea advanced more than 1000 km, forcing any people living there to abandon their settlements. This advance would have been most rapid during the second half of the period of sea-level rise, after the Younger Dryas.

The most obvious form of erratic behaviour would have been the sudden advance of the sea as a result of a storm surge. During the inexorable rise, the waters at times flooded across the flattest parts of the Persian Gulf at about a kilometre per year. This would have meant that during a major storm the surge in sea level might have devastated land several kilometres in from the temporary shoreline. Repeated events of this nature could easily have been remembered in terms of a single catastrophic event rather than long-term change.

The second potential source of the Flood Myth in this particular region is that there were times when rising sea levels put on a sudden spurt. More dramatic accelerations would have occurred when there were major meltwater releases from the freshwater lakes that formed behind the melting ice sheets. The extreme case was the 163 000 km^3 outburst from Lake Agassiz at 8.2 kya (see Section 2.9) (Leverington, Mann & Teller, 2002). Instead of the steady advance of the sea of a few metres per year, this event, which might have taken place in a year or so, could have caused an advance of more than 10 km across the flattest parts of the Persian Gulf.

There is one other consequence of rising sea level that could have led to the sudden inundation of large areas of a low-lying land. This is where an inland sea is cut off from global changes until the sea rises to a given level. The best candidate for this is the Black Sea. This proposal has received widespread attention thanks to the work of Walter Pitman and William Ryan (Pitman & Ryan, 1997; Ryan *et al.*, 1997). They have presented evidence that at the end of the last ice age the Black Sea was a freshwater lake cut off from the salty Mediterranean, and its level was around 150 m lower than now. As the global sea level rose, the narrow strip of land separating the Black Sea from the Mediterranean was breached around 8.4 kya. The timing of this event may be related to the sudden rise in sea level associated with the draining of Lake Agassiz around this time. Within a few years more than 100 000 km^2 of land around the Black Sea was swallowed up by the massive influx of water. The people who escaped from this locally devastating event would have carried with them a tale of epic proportions that could easily have been perpetuated throughout the Middle East as a Flood Myth.

The Black Sea flood has now become a matter of intense scientific debate. A group headed by Ali Aksu, at the Memorial University of Newfoundland, argue that cores taken from the Marmara Sea tell a different story (Aksu *et al.*, 2002). These suggest that the Black Sea rose to reach the Bosphorus sill between 11 and 10 kya, at which time low-salinity water started to flow out through the Marmara Sea and into the Mediterranean. By around 9 kya the global sea level rose to match that of the Black Sea. Thereafter, there was a two-way flow through the Bosphorus with low-salinity water flowing out in the upper layer and more saline water flowing into the Black Sea at lower level. This pattern would then explain the appearance of salt-water-tolerant molluscs around 8 kya: not the product of a massive inrush of salt water, but the eventual establishment of a balanced condition. If this interpretation is correct then the Black Sea flood is a case of, in the words of Thomas Huxley, 'the slaying of a beautiful hypothesis by an ugly fact.'

Recent modelling work by Mark Siddall, formerly of the Southampton Oceanography Centre, UK, and co-workers at the Woods Hole Oceanography Institute, MA, USA (Schiermeier, 2004) indicates, however, that features of the topography of the floor of the Black Sea are consistent with a major inflow around 8.5 kya. Although this work implies that the flood may have taken several decades to raise the level of the Sea, it supports the idea that the biblical Flood Myth could be connected with events in the Black Sea in the early Holocene. But, unless unequivocal evidence is found of settlements dating from before 8.4 kya, and lying well below current water levels, this is another debate that looks set to run and run.

An even more localised explanation for the widespread nature of the Flood Myth is that it reflects local experience. With the coming of agriculture many settlements were on the flood plains of rivers. These areas offered obvious opportunities, including the most fertile and easily tilled soil, plus a plentiful and reliable source of water and scope for irrigation. The price to pay for these benefits was the risk of flooding. While normal fluctuations in seasonal inundations were something that any community would have to take in its stride, the occasional extreme flood that might only occur once in, say, a century was another matter. The survivors of such a catastrophe would not only regard themselves as having been smiled upon by the Gods, but also have elevated the event to a mythical status. It is hardly surprising that Gilgamesh, the oldest Flood Myth, hails from Mesopotamia. Sudden changes in the course of the major rivers would have had catastrophic consequences. There is also abundant archaeological evidence that occupation of cities was interrupted by rare but massive inundations that buried everything in deep layers of silt. Conversely, in the Nile Valley, where the inundations were such a regular feature of life, there is no equivalent of the Flood Myth.

5.16 THE FORMATION OF THE NILE DELTA

The consequences of rising sea levels are not simply the inundation of low-lying coastal areas. In some places they altered the estuarial aspects of major rivers and produced rich alluvial deltas. This is what happened around 6 to 7 kya in the Mediterranean where the rising sea altered the drainage characteristics of the Nile Delta. Cores from the delta provide a detailed picture of how the alluvial deposits built up (Krom *et al.*, 2002). By 6.5 kya the level was about 15 m below current values. Up until this time, the Nile sloped steeply down to the sea, some 80 km farther north than the current shoreline, and the silt was swept out to sea. As the rise in sea level slowed, the conditions were right for the sediment from the Nile to build up and form a much more extensive rich alluvial fan. At the same time, the flow rates in the Nile were declining from the high levels in the early Holocene, when they were triple the modern values (before the Aswan Dam was built). Although the first evidence of farming in the delta region dates from about 8 kya, the growth of the delta played a major part in establishing the agricultural wealth of Egypt. Agriculture in the Nile valley itself did not take off until around 5.5 kya, possibly as a consequence of desertification in the eastern Sahara (see Section 5.17).

Detailed analysis of the sediment in Nile delta cores also provides an accurate record of changes in Nile river flow over the past 7 kyr. In particular, studies of the strontium isotope ratio of the sediment can identify whether the principal source of the floodwaters is the Blue Nile, originating in the Ethiopian Highlands, or the White Nile, which originates farther south in Lake Victoria on the Equator. Prior to 6 kya the flow rate was much higher and there was a lower proportion of Blue Nile particulates. This suggests that times of high flow more rain fell in the Ethiopian Highlands as a result of the northward movement of the ITCZ. The records also confirm a marked decline in flow rate of the Nile between 4.7 and 4.2 kya. During this period the flow from the Blue Nile formed a higher proportion of the flow. From 4.2 to 3.1 kya, the flow increased, but with greater fluctuations in discharge from decade to decade. Flood level decreased until around a thousand years ago, since when the flow has stabilised at a relatively low level.

5.17 THE LOST SAHARAN PASTORAL IDYLL

Changes in rainfall over North Africa in the mid-Holocene had a drastic impact on life in the Sahara. The rock and cave paintings in the central Sahara, the best known of which are the 'Tassili Frescoes', depict a lost world of a land teeming with life and supporting vibrant pastoral economy.[5] These images of antelope, elephant, giraffe, hippopotamus, ostrich, rhinoceros and pastoral lifestyle seem impossible when viewed against the surrounding superarid desert (Fig. 5.4). The climatic record is, however, unequivocal: for most of the time between around 14 kya and 5 kya the Sahara experienced a monsoonal climate (see Fig. 2.11). The region had considerably greater rainfall than now and much of the land had permanent vegetation. The suddenness of the changes, both in the onset of the wetter conditions during the Bølling/Allerød period, and similarly the rapid onset of drier conditions in the mid-Holocene, are a fascinating example of the non-linearity of the climate (see Section 2.11). They also provide an interesting example of the changes humans had to adapt to during the supposedly stable climate of the Holocene, especially in coming to terms with the desiccation of the Sahara.

Following the Younger Dryas, when the general cooling of the climate around the North Atlantic brought drier conditions to the Sahara, the summer monsoon spread much farther north. From the Horn of Africa in the east to Mauritania in the west vegetation zones moved northwards with the summer rains. Savannah woodland and open dry forests grew in regions that are now hyperarid desert.[6] By 8 kya Lake Chad covered a vast central area of some 330 000 km^2 (Fig. 5.5) To the east of the lake there were savannah grasslands that were periodically inundated. Rivers ran out of the central mountain areas of the Hoggar, Tibetsi and Darfur to fill this vast lake. Rivers also ran eastwards along what are now Wadi Howar and Wadi Melik to form tributaries of the Nile, and southwards into the Niger. Elsewhere large lakes developed, as can easily be seen on the ground by their characteristic sedimentary deposits and fossil shorelines. The

[5] See the Tassili frescoes website http://www.paleologos.com/saharaan.htm.

[6] The data presented here draw on the extensive work done by the University of Mannheim on palaeovegetation maps of Africa, details of which can be found on http://www.uni-mannheim.de/phygeo/palaeo.htm.

FIGURE 5.4. An example of a rock engraving (petroglyph) from the central Sahara showing the exotic wildlife that existed there during the early Holocene.

impressive river systems that drained into these lakes have been detected beneath the shifting sands by ground surveys and mapped by radar imagery from space (Fig. 5.6). These rivers not only drained eastwards into the Nile, but also flowed northwards and discharged directly into the Mediterranean.

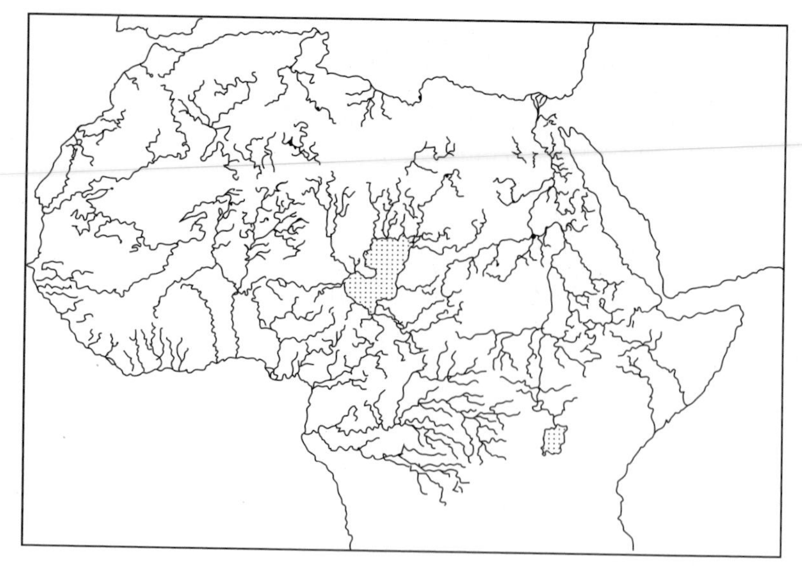

FIGURE 5.5. Map of northern Africa showing the more extensive river systems around the Sahara and greatly expanded Lake Chad.

It appears that the first incursions of nomads into the southern Sahara, which came from the south, did not take place until around 12 kya. At around the same time migrants may also have entered the northern Sahara from the Mediterranean coast. There is evidence of their presence by 11.5 kya in the Acacus Mountains of Libyan Sahara. What is even more interesting is that between 9 and 8 kya these people had developed a hunting strategy that involved the capture, penning and feeding of Barbary sheep to manage their food supplies more efficiently (di Lernia, 2001). These early arrivals were reinforced by a new influx from the east around 8 kya. This wave of migration appears to have brought a pastoral economy to the region with the introduction of domesticated animals.

Attempts to use genetic mapping to shed light on migrations across the Sahara have focused on movements up and down the Nile Valley. These show that such movements have a complicated history. The basic observation is that the prevalence of a genetic marker associated with populations in sub-Saharan Africa (see Section 3.8)

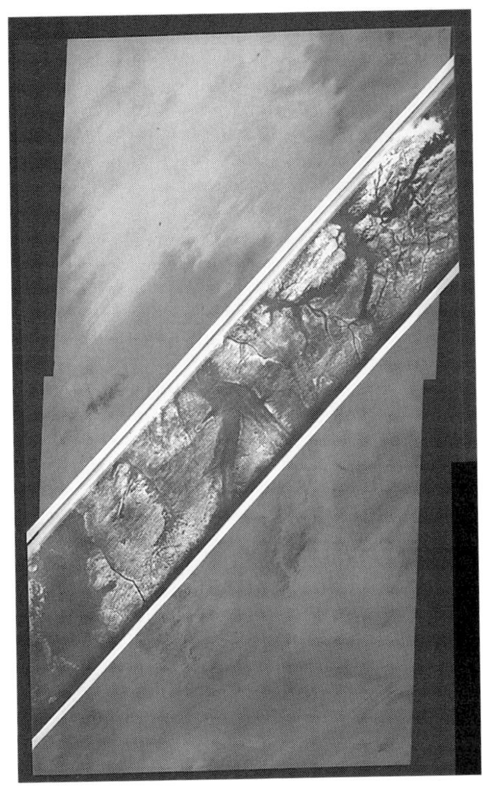

FIGURE 5.6. Satellite radar image of part of the Sahara showing the extensive drainage system that existed in the early Holocene, superimposed on a photographic image of the surrounding terrain, which is now covered by drifting sand.

falls from a high level in southern Sudan to a much lower level in Egypt (Krings *et al.*, 1999). Furthermore, studies of the remains of ancient Nubians show that the proportion of this marker was the same then as it is for people now living in the area. This indicates that migrations had occurred in both directions along the Nile, which is consistent with historical evidence for long-term interactions between Egypt, Nubia and farther south. The bulk of these migrations probably occurred within recorded history and the flow from north to south was either earlier or less extensive than that from south to north.

There is no doubt about the sudden decline in rainfall across the Sahara. Around 5.5 kya deep-sea sediment records off the west coast of North Africa show a sudden increase in the amount of dust being transported in the winds from the east (see Fig. 2.11) (deMenocal *et al.*, 2000). At the drop of a hat the climate shifted to a drier form and the desert began to take over. A survey of conditions around 5 kya shows a general drying, although the distribution and timing of these changes show interesting regional variations. In particular, the eastern Sahara aridification, which may have started as early as 6.5 kya, was already well advanced with the Libyan desert extending southwards into what is now Sudan. Only in favoured wadis did the savannah remain a feature of regions that are now desert.

Computer model studies of the early and mid-Holocene can produce simulations that reflect this pattern of changes. Temperatures were slightly lower than at present, in part because of increased cloud cover and more humid conditions. The models show increased monsoonal precipitation in the eastern Sahara and East Africa (Kutzbach & Liu, 1997; Bonfils *et al.*, 2001). Simulations of vegetation cover show an upward shift of montane vegetation belts, fragmentation of drought-tolerant vegetation in southern Africa, and a major northward shift of the southern margin of the eastern Sahara. As noted above the existence of large lakes confirms the prediction of wetter conditions across northern Africa. The lake and pollen data also confirm the northward expansion of the monsoon in eastern Africa, but the model is unable to simulate the wet conditions in western Africa.

As for human activities, the early decline in rainfall in the eastern Sahara may have led to the oldest-known astronomical alignment of megaliths in the world (Wendorf & Schild, 1998). Consisting of a set of huge stone slabs in the desert of southern Egypt, known as Nabta, it forms a stone circle, a series of flat, tomb-like stone structures and five lines of standing and toppled megaliths, and is dated at around 6.5 to 6 kya. The stone slabs, some of which are 3 m high, were dragged more than a kilometre to the site. They lie on the shoreline of an ancient lake, which was used by nomads for grazing their cattle during the summer rainy season. To judge from carbon dating of charcoal and ostrich shells, occupation may have started as early as 12 kya. Initially the settlements at Nabta were small seasonal camps of cattle-herding and ceramic-using people. These cattle are regarded as the first example of the African pattern of herding. Cattle served as a 'walking larder' and provided milk and blood, rather than meat, as they do to this day for the Masai in Kenya. The site was intensively exploited by around 9 kya. This included the digging of walk-in wells that suggest the site could have been occupied throughout the year. Nomads used the area until about 5.6 kya, at which time it became hyperarid and uninhabitable.

Nabta is close to the Tropic of Cancer, and so the noon sun is vertically overhead around the summer solstice. The vertical sighting stones in the circle form a calender system, when combined with an east–west alignment between one megalithic structure and two stone megaliths over a kilometre distant, and two other geometric lines involving about a dozen additional stone monuments that lead both northeast and southeast from the same megalith. The timing of the construction of the array coincides with the onset of increasingly erratic rainfall patterns. It is possible that the Nabta people were reacting to these changes by using the array to measure the summer solstice and predict the onset of the rainy season. The megalith may have been part of a variety of ritual activities as there were a number of tumuli containing the bones of cattle on the site.

There may be a link between people at Nabta and the Neolithic people living along the Nile. This could have contributed to the rise of social complexity in ancient Egypt. Clearly the people at Nabta had a structured society that enabled them to plan the arrangement of their villages, the excavation of large, deep wells, and the construction of complex stone structures made of large shaped and unshaped stones. Other Nabta features, which subsequently appear suddenly and without evident local antecedents in the late Predynastic and early Old Kingdom in the Nile Valley, are the role of cattle to express differences of wealth, power and authority, the emphasis on cattle in religious beliefs, and the use of astronomical knowledge and devices to predict solar events. Many of these had been well developed for a long time at Nabta.

It may be that as the desert became more and more inhospitable the crocodile-infested swamps of the Nile grew more attractive. Indeed, as the monsoon ceased to spread so far north the flow of the lower Nile may have abated somewhat. Whatever the changes, it is an interesting fact that the earliest records of Egyptian history relate to the period around 5.15 kya (3150 BC) that is only a century or two after the Sahara started its descent into the formidable desert we now know.[7] So the legend of the Scorpion King in Egyptian prehistory may be a tribal memory of an earlier pastoral savannah existence.

More relevant here, these changes are a natural bridge between the impact of climate change on prehistory and the recorded histories of the ancient civilisations. Saharan desiccation could have led to the rise of Dynastic civilisation in Egypt. Perhaps the nomads, who built the megalithic structure at Nabta, brought their more complex cosmology to enrich the culture of the Nile. More directly, population pressures in the south, and the collapse of the pastoralism, could have led to a move to agriculture as people that had previously practised seasonal migration settled permanently in the Nile Valley. This move around 5.5 kya may have stimulated both technological innovation and social stratification in Predynastic Upper Egypt. In time, this may have been the stimulus for the unification of Upper and Lower Egypt around 5 kya and the creation of the Old Kingdom.

In the central and western Sahara the tale of desiccation emerges somewhat later. The record from the deposits formed in the Kajemarum Oasis, in the Manga grasslands of northeastern Nigeria, spanning the last 5.5 kya, reveals the episodic deterioration of the climate in the region to the south of the Sahara (the Sahel) (Street-

[7] Up to this point all dates have been expressed in units of kya (thousand years ago), but once we get to dates that are more commonly expressed in terms of BC and AD, it is easier for the reader either to have both forms presented or whichever form appears to be the most appropriate.

Perrott *et al.*, 2000). Desert-dust deposition began to increase around 4.7 kya. Rainfall during the summer-monsoon season declined permanently after 4.1 kya. While the presence of groundwater maintained vegetation in some wadis for a considerable period of time, the inexorable desiccation changed the landscape. Farther west, however, there is evidence that the moister conditions continued longer with only a slow transition towards current conditions.

These changes in the Sahara did not mean that all human activities came to an abrupt end. For example, the Wadi Tanezzuft, which originates in the Tassili and Acacus mountains before disappearing 300 km to the north beneath the dunes of the Ubari erg, had abundant water and fed several lakes along its course. Pastoral communities intensively occupied it during the mid-Holocene. With the coming of the drought, the local consequence was a decline in the wet areas. Lakes and swamps between the sand dunes of the ergs dried up quickly, but the main course of the Wadi was still active between 3 and 2 kya. The increasing aridity is, however, evident from the progressive retreat of archaeological sites to the south.

The onset of the present superarid conditions did not occur until around 1.5 kya. Before this final decline came the blossoming of the kingdom of the Garamantes that supplied salt to West Africa (Cremaschi & di Lernia, 1999). This kingdom in the oasis of the Wadi Tanezzuft appears to have benefited from a brief period of increased rainfall between 2.45 and 2.25 kya. Information about precipitation changes has been obtained from a dendroclimatological study of wood from trees of the Cypress family (*Cupressus dupreziana*) which had been imported from the Wadi Tanezzuft and used to make house doors in the old cities of Ghat and Barkat, Libya.

A progressive decline of rainfall from 2.25 kya corresponds to a period when the Garamantian kingdom tried to react against increased aridity by intensifying horticulture, establishing walled citadels and developing the long-distance trade across the Sahara using camels. This delayed the final collapse of the kingdom until around 1.6 kya. The onset of hyperarid conditions coincided with a sharp decrease in tree-ring widths in the dendroclimatological measurements. Thereafter, the only use of Wadi Tanezzuft was as part of the main caravan route connecting the Mediterranean coast to the central Africa.

5.18 THE BANTU EXPANSION

Another aspect of the desiccation of the Sahara appears to have been the movement of people southwards. This pressure may have precipitated what is often termed the 'Bantu expansion'. The term Bantu, which means 'people', covers some 700 languages spoken by 200 million people in sub-Saharan Africa. Sir Harry Johnston in the late nineteenth century postulated that the languages had spread from a relatively small group of ancestors in the present-day Cameroon–Nigeria region. Subsequent work in the 1950s supported this hypothesis, while limited archaeological evidence suggested that the expansion started as recently as 5 to 3 kya, and there is now a consensus in support of this timescale. Genetic Y-chromosome studies (Cruciani *et al.*, 2002) show that, of the sub-Saharan populations, Bantu populations have the lowest level of internal genetic diversity. This suggests a recent dispersal of these populations dated at around 4 kya. Furthermore, although the Bantu clearly stand out as a separate group, they show clear genetic relations with West Africans. The genetic studies thus support the archaeological consensus concerning the timing of this movement.

The move southwards has been attributed to a variety of causes. These include environmental and population pressures. The widespread evidence of increasing aridity in both the Sahel region and equatorial Africa, associated with the event around 4.2 kya suggests that this climatic deterioration may well have been the trigger for the Bantu expansion. Economic and social factors relating to the development of agriculture and the spread of iron metallurgy were probably as important for the successful takeover of southern Africa as the changing climate that forced them out. The ability of the Bantu to grow millet and sorghum not only ensured a good diet, but also encouraged the establishment of more stable settlements in fertile regions. They multiplied faster and were better fed than their Stone-Age neighbours. Their use of iron enabled them to manufacture tools and weapons. The tools improved the efficiency of their farming and their weapons made their armies invincible.

The actual movement may have occurred in a series of waves across the rainforest to the south using the major waterways, or possibly around the edge of it. When it comes to the details of this large-scale movement, there are, however, considerable uncertainties. Many of these centre on the evolution of the various Bantu languages. Nevertheless, there can be little doubt that the drying out of the Sahel, around 4.2 kya, would have reduced the carrying capacity of the region and driven people southwards. Where adequate water supplies remained people continued to thrive. For example, around Lake Chad conditions were sufficient to sustain a vibrant social structure throughout historic times.

Genetic mapping of African cattle provides additional insight into the origins and movements of indigenous breeds. Because wild cattle are not native to sub-Saharan Africa it is possible to explore the origins of the present cattle distribution in Africa (Hanotte *et al.*, 2002). The earliest African cattle appear to have been domesticated within the continent, possibly in the eastern Sahara, around 10 kya, but genetically influenced by cattle domestication in the Near East and the Indus Valley. They then expanded from the region of origin to reach the southern part of the continent by following an eastern route rather than a western one, which suggests that the Bantu expansion may have followed this route.

5.19 ENSO COMES AND ENSO GOES

Up to this point the prehistory of Central and South America has been kept on one side. The reason for this selective approach is that any analysis of the impact of climate change on human societies in these parts of the world is dominated by the behaviour of the Pacific Ocean. Here what matters is a quasi-cyclic phenomenon resulting from ocean– atmosphere coupling across the Pacific and known as

El Niño / Southern Oscillation (ENSO) (see Section 2.8), which has shown substantial shifts in its behaviour during the Holocene. For much of the early Holocene there was relatively little ENSO activity, then around 5.5 kya El Niño events returned (Tudhope *et al.*, 2001). This increase in activity may have been part of wider climatic changes that led to the shift in the central and eastern Sahara, plus Arabia, to extreme aridity within a century of so.

The apparent absence of ENSO activity in the early Holocene is odd. It has been assumed that recent record-breaking El Niño events in 1982/83 and 1997/98 were a symptom of global warming. On this basis, it might be expected that during the greater warmth of the early Holocene these events would be more common. The evidence is, however, unequivocal. The most comprehensive set of data comes from cores taken from Laguna Pallcochoa in southern Ecuador (Rodbell *et al.*, 1999; Sandweiss, Maasch & Anderson, 2001). These clearly show that, whereas in the early Holocene there was a marked absence of ENSO events, things changed in the mid-Holocene. The frequency of events first became significant around 7 kya, with the onset of the modern quasi-periodic behaviour becoming dominant around 5 kya. The incidence of increased variability was not monotonic, but came in pulses every 2 kyr or so, reaching a peak around 1.2 kya, since when it has declined sharply. The millennial periodicity may be related to a similar fluctuation associated with Dansgaard/Oeschger (DO) events during the ice age (see Section 2.5) that had a period of around 1.5 kyr. This periodicity appears to have continued in the North Atlantic during the Holocene in a much less pronounced form with a period around 1.8 kyr.

The changing frequency of ENSO events appears to have exerted a major influence on the prehistoric civilisations of South America. The many river valleys supplied by the meltwaters of the high Andes that cut through the arid coastal plain of Peru were the sites where prehistoric societies built monumental architecture. The combination of the exceptionally rich fisheries along the coast and irrigation-based agriculture was capable of supporting a large population. Dating of various major constructions has identified the times when these massive edifices were built. The earliest of these, the city of pyramids known as Caral, has been studied by the archaeologist Ruth Shady Solis, from the University of San Marcos, Lima, Peru (Solis, Haas & Creamer, 2001). Occupation of the site dates from around 4.9 kya: much older than anything else in South America. This elaborate complex of pyramids, temples, an amphitheatre and ordinary houses appears to have been occupied for about a thousand years.

Caral appears to have represented the pinnacle of the civilisation that existed during the late third millennium BC. Crucially, there is not the faintest trace of warfare at Caral: no battlements, no weapons and no mutilated bodies. Instead, the archaeological evidence suggests it was a gentle society, built on commerce and pleasure. Beautiful flutes made from condor and pelican bones were found in one of the pyramids. There is also evidence of a culture that took drugs and perhaps aphrodisiacs.

It may be no coincidence that the really significant surge in ENSO activity began around 5.7 kya and reached a peak at 4.9 kya. This would have led to increased rainfall in Peru, especially in the Andes, during El Niño conditions. Although this activity only lasted a few hundred years, once Caral had become established it may have been able to survive the subsequent drier centuries before eventually being abandoned and left untouched beneath the sand of the Peruvian desert.

The next pulse of ENSO activity began during the second millennium BC. From then onwards, until relatively recent times, activity remained at a higher level, although there were significant fluctuations in level over the centuries. This renewed activity, which ran from around 3.5 to 2.4 kya, seems to have stimulated the next upsurge in social structures that led to the widespread building of monumental architecture.

6 Recorded history

> 'My name is Ozymandias, king of kings
> Look on my works, ye Mighty, and despair!'
> Nothing beside remains. Round the decay
> Of that colossal wreck, boundless and bare
> The lone and level sands stretch far away.
>
> Percy Bysshe Shelley (1792–1822) *Ozymandias*

Moving on to recorded history takes us beyond the simple matter of looking for instances of disruptive climate change that appear to have influenced the pace of human development. We now have the parallel written record that may contain valuable support for theories of the impact of climate change on the fortunes of early civilisations. This includes texts, inscriptions on monuments and stele, which could contain direct references to adverse conditions (e.g. drought, harvest failures and famine). More frequently, we have to rely on indirect evidence in the form of the dating of major events that marked the downfall of civilisations and the cessation of written records. Often these collapses are identified by the abandonment of cities and towns in terms of artefacts found on the sites.

An additional feature of this complementary analysis is that it seeks to explain the rise of ancient civilisations in terms of their particular resource advantages. Where climatic factors loom large, as in the case of Mesopotamia and Egypt, this complementary approach is central to the whole analysis. The first step is to establish for these civilisations, where climatic circumstances were particularly auspicious, the criteria for success. Then it is easier to identify where climate change played a part in the downfall of societies whose climatic foundations were more precarious. This analysis can then be extended to explore how some of the more social aspects of our lives may have been affected by the impact of climate change on the development of early civilisations. In this way it will explore various aspects of our belief systems, gender roles, public health and relationships with animals that have been touched on in earlier chapters.

This approach is bound to be selective. The aim is to consider how the issues explored in the earlier chapters elide into recorded history. In practice this means overlaying our climate change template with what is known about the history of early civilisations. In addition, this process will draw on a wide variety of climatic sources, of which the most important are recent efforts to collate available data to provide a coherent picture of conditions at around 6 kya.[1] The process needs to tie in with the data on the global periods of rapid climate change identified by the CASTINE project (Climatic Assessment of Transient Instabilities in the Natural Environment; see Section 2.7), notably between 3.5 and 2.5 kya, together with the period of more regional change around 4.2 to 3.8 kya. This approach allows us to concentrate on aspects of the climate record that may shed some light on those upheavals where the available historical records do not provide much evidence of the role of climate change. But, in so doing, it helps to have an idea of the climatic conditions that prevailed at the time in those regions where ancient civilisations became established.

6.1 CLIMATIC CONDITIONS IN EUROPE DURING THE MID-HOLOCENE

As noted in Chapter 5, we can combine various proxy records of the climate during the Holocene together with our knowledge of the North Atlantic Oscillation (NAO) to improve our understanding of the climate across Europe and the Middle East. At the time of the emergence of the first major civilisations in southwest Asia between 5.5 and 5 kya the climate in the northern hemisphere was nearing the

[1] This account of climate change during the Holocene draws on a variety of standard texts on the subject (e.g. those by Lamb and by Grove; see Bibliography), and major coordinated studies such as BIOME 6000 (Prentice *et al.*, 2000), which provide a combination of time series throughout the Holocene and snapshot's at specific times (e.g. around 6 kya).

end of the Holocene optimum. The climatic shifts that had had such a major impact on the Sahara had yet to lead to significant changes farther north.

Pollen data from across Europe (see Section 5.2) suggests that, around 6 kya, winters were 1 to 3 °C warmer than during the twentieth century in the far north and northeast of Europe, but 2 to 4 °C cooler in the Mediterranean region. At the same time, summers were warmer then in northwest Europe and in the Alps, but cooler than present at lower elevations in southern Europe. Possibly more important, summers were drier in northwest Europe and the Alps, but moister in southern and eastern Europe. This is consistent with the conclusion that the climatic patterns in mid-latitudes of the northern hemisphere at this time – identified as the Atlantic period – featured stronger westerlies. This pattern coincides with the positive phase of the NAO that, in particular, produces cooler wetter winters in the Middle East.

Around 5.5 kya there is evidence of a general cooling in mid-latitudes of the northern hemisphere, which does not show up markedly on the pollen data. This is often referred to as the *Neoglaciation* as it coincided with an expansion of the glaciers in the Alps and Scandinavia. This seems to have been linked with the more wide-ranging changes that were occurring in the tropics. A famous individual consequence of this cooling appears to have been the burial in ice of the prehistoric mummified corpse known as the 'Iceman', at the upper edge of the alpine glacier at the head of the Ötzal. His death is dated as being around 5.3 kya. It would not be until the end of the twentieth century that conditions were warm enough to melt his icy tomb and reveal his body.

After the Neoglaciation the climate in Europe recovered some degree of warmth, but remained relatively dry, consistent with the definition of the sub-Boreal climate. The evidence for a significant regional change around 4.2 kya is, however, limited. In terms of the North Atlantic and Siberia the circulation was weak. European glaciers were stable or in retreat, and the treeline rose in Sweden.

Thereafter, the broad picture is of a gradual move to a cooler wetter climate. The most important features of this trend were a cooler period between 3.5 and 2.5 kya, followed by warmer drier conditions until around 1.5 kya (AD 500), then a colder spell until about AD 1000. During the last millennium the important features have been the relative warmth of the eleventh and twelfth centuries (the Medieval Climatic Optimum) and the colder conditions from the fourteenth to nineteenth centuries (the Little Ice Age), but these ups and downs fall outside the scope of this book.

6.2 EAST ASIA IN THE MID-HOLOCENE

Conditions across much of China at 5 kya seem to have been warmer than present, but perhaps cooler than in the early Holocene. These conditions seem to have remained similar to those at 8 kya, with warm temperate forest extending hundreds of kilometres further north than at present. Over much of China, pollen records indicate temperatures 2–4 °C warmer than at present (perhaps as much as 5 °C higher in the Tibetan Plateau), cooling after about 4 to 3 kya. In northwestern China increased dust deposition indicates that the climate became much more arid from 6 to 5 kya. This dry, dusty period interrupted the formation of the brown soil developing under a warm-humid subtropical climate during the Holocene Optimum. It is also possible that this aridity led to the marked decline in Neolithic culture in the periods of the middle and late Yangshao period (6–5 kya) in the Guanzhong Basin (Yu *et al.*, 2000).

In the northeast of China (Manchuria), peat deposition seems to have begun in the mid-Holocene, coincident with a cooling of climate after 5 kya. Lake levels indicate conditions moister than present over most of China up until 3.5 kya. Magnetic susceptibility of loess profiles presents the same picture. Evidence of Neolithic agriculture in northwestern regions of China that are currently too arid for crop-growing is further testimony of the moister climate that prevailed at around 5 kya. Agriculture was already present and expanding throughout the southeast Asian region, but deforestation in southern China and in the monsoon zones of Indo-China does not appear to have been significant until after around 4 kya.

6.3 AGRICULTURAL PRODUCTIVITY: THE ABUNDANCE OF MESOPOTAMIA

Although the changes described in the last two sections applied to many places where civilisations were emerging, in the case of Mesopotamia and Egypt, there are more basic climatic issues to address first. In the case of southern Mesopotamia, when confronted in summer by its blisteringly hot, arid plains, archaeologists have to address a basic question: why did the first cities appear here? Covering an area of some 30 000 km^2, the flat, river-made land of Sumer had no minerals, almost no stone and no trees. In summer the daytime maximum temperatures average around 40 °C and often reach 50 °C. Annual rainfall is about 150 mm and it is bone dry for eight months of the year. Winter nights are cold, and the strong north winds can bring squally rainstorms. In spring, the melting snows of the Taurus and Zagros mountains produce flash floods. In bad years, these swept everything before them. While the climate of the mid-Holocene may have been moister, these fundamental climatic challenges were part of the development of the region. Yet by around 5.8 kya as many as 10 000 people may have lived in the city of Uruk.

The key to the success of southern Mesopotamia was the harnessing of its massive agricultural potential through irrigation. The rich alluvial soil of the region was able to produce huge yields provided it was adequately watered. So, the emergence of more organised city states in this region before anywhere else in the world was driven by the need for large-scale public works to overcome the climatic challenge of the area and exploit its agricultural potential. How this occurred was, however, closely linked to the changing climate and geography of the time.

As discussed in Section 5.15, by about 8.5 kya, rising sea levels had covered much of the present-day northern Persian Gulf. A large estuary formed where the Euphrates River exists today. As the sea level rise slowed around 7 kya, the river estuaries receded from their northernmost limits, which extended along the Euphrates as far north as Ur.[2] Then the estuary began to fill with silt, impeding the natural drainage and forming large swampy areas, and the climate became more arid. These changes took place as the first farming communities appeared. As the climate became drier, communities began small-scale irrigation farming. Crop yields shot up.

From around 7 kya the combination of well-watered land and high water tables had created favourable conditions for irrigation agriculture. At first this process would have been piecemeal, but over the centuries it would have become more organised. As farmers gained better access to river water, they inundated more extensive canal networks. The intelligent use of water supplies yielded rich dividends. Community leaders who organised canal digging, and the maintenance of waterworks and dams, were often seen as possessing supernatural powers, and to be capable of interceding with the gods. The deities controlled the hostile forces of nature: the bitter winter north winds; spring floods; and summer heat waves that withered crops in the fields. In time, these people became the spiritual and political leaders. Their role became more ritualistic, involving the giving of offerings to the gods at small village shrines erected on sacred ground.

Archaeological evidence suggests that this process may have started as early as 7.8 kya, and by around 6.3 kya there were several large centres of population. This lengthy period of development is usually known as the 'Ubaid period'. As the communities grew the scale of these activities blossomed and the religious monuments grew ever bigger. Indeed, the growth in population, fuelled by the productivity of the local agriculture, may well have led to the need to generate public works to exploit surplus labour. Although there is some

[2] There are various theories about the marshlands at the mouths of the Tigris and Euphrates being the origin of the Garden of Eden myth. Depending on the timing of any human occupation, it is interesting to speculate whether it site should be identified with these marshlands or land that was inundated as the sea level rose in the Persian Gulf.

evidence of drier conditions in Mesopotamia around 5.5 kya (see below), this may have increased the reliance on irrigation and stimulated the move to bigger cities. By around 5.2 kya the city of Uruk had grown to some 50 000 people, who lived behind substantial defensive walls. The period of growth from around 6.3 kya to around 5.1 kya is symbolised by the expansion of Uruk and is usually named after this city.

The motor for these developments was the incredible fertility of the alluvial soil between the Tigris and Euphrates rivers. Cuneiform tablets from the city of Ur, dating from around 4.1 kya, suggest that yield-to-seed ratios of 30:1 were normal for irrigated fields in southern Mesopotamia, and that higher ratios (up to 50:1) were possible. Inevitably, such figures have to be treated with caution, but they appear to explain why the region could sustain large populations living in substantial cities filled with such magnificent ritual edifices. Herodotus, regarded as being an extremely reliable source of information, has quoted even greater figures. Although he was writing nearly two millennia after the earlier records, his observations of the productivity of grain production, farther north in Assyria, are striking. He notes:

> But little rain falls in Assyria, enough, however, to make the corn begin to sprout, after which the plant is nourished and the ears formed by means of irrigation from the river. For the river does not, as in Egypt, overflow the corn-lands of its own accord, but is spread over them by the hand, or by the help of engines. The whole of Babylonia is, like Egypt, intersected with canals. The largest of them all, which runs towards the winter sun, and is impassable except in boats, is carried from the Euphrates into another stream, called the Tigris, the river upon which the town of Nineveh formerly stood. Of all the countries that we know there is none which is so fruitful in grain. It makes no pretension indeed of growing the fig, the olive, the vine, or any other tree of the kind; but in grain it is so fruitful as to yield commonly two-hundred-fold, and when the production is the greatest, even three-hundred-fold. The blade of the wheat-plant and barley-plant is often four fingers in breadth. As for

> the millet and the sesame, I shall not say to what height they grow, though within my own knowledge; for I am not ignorant that what I have already written concerning the fruitfulness of Babylonia must seem incredible to those who have never visited the country.

These figures are phenomenal by comparison with any values from pre-industrial Europe or North America. As noted in Chapter 3, the figures for medieval England had famine levels of only two grains for every grain sown, while abundance was when productivity rose to five or six grains for every one sown. Perhaps more important is that, in areas that were once Mesopotamia, modern farms, which do not use irrigation, produce yields of seven to nine seeds to one sown. This suggests that the benefit of irrigation was at least a threefold increase in yield.

The importance of these figures is not so much in their absolute values, but in the scale of the advantage of irrigation. The combination of the rich alluvium of Mesopotamia and the reliable water supply of the Tigris and Euphrates provided the ideal mixture for producing abundant harvests. The consequent surpluses would have been the obvious springboard for a series of social developments. So there is no doubt about the fertility of southern Mesopotamia and the incentives to exploit its riches. What is less clear is whether, as part of this exploitation process, climatic changes played a part in the move to establishing cities in the region.

Recent results obtained from a speleothem in Oman (Fleitmann *et al.*, 2003) provide a record of the summer monsoon (July to September) between 10.3 and 2.7 kya and between 1.4 and 0.4 kya. This shows a marked long-term decline from high levels between 10 and 8 kya, including a notably dry period between 5.5 and 5 kya. In addition, shorter-term shifts, notably around 8.2 kya, suggest that the northward movement of the monsoon rains in the early Holocene was governed by the glacial boundary conditions in higher northern latitudes. Thereafter, the extent of the monsoon was controlled by the changes in solar insolation that had such a profound effect on the Sahara (see Section 5.15). To the extent that these summer rains extended into southern Mesopotamia and affected the initial establishment of agriculture in the region, the decline in monsoon rains could have played an important part in the development of irrigation here.

The other recent source of detailed climatic information comes from the analysis of sediment cores from Lake Van in Eastern Anatolia (Wick, Lemcke & Sturm, 2003). This work covers the pollen content of 13 kyr of annual layers (*varves*) and shows that the optimum climatic period for vegetation, marked by higher rainfall, was between 6 and 4 kya. This suggests that during this period the amount of rain and snow in the Anatolian highlands would have been greater than at present and a reliable source of both winter run-off (December to March) and spring snowmelt (April to June) to the Tigris and Euphrates basin. So, for southern Mesopotamia, while local rainfall may have declined, especially around 5.5 kya, the supplies for irrigation remained plentiful.

The combination of changes in rainfall in the period 6 to 4 kya would have had interesting consequences for southern Mesopotamia. The higher run-off in the December–June period would have increased the propensity for rivers to change their courses and would have been an incentive to manage floods and exploit the water more effectively. At the same time, increasing summer aridity would have increased the sensitivity of agriculture to drought. These pressures may well have been the cause of increasing social complexity and organised structures to respond to challenges posed by climatic change. In addition, the greater emphasis on public works and organised community efforts may have altered gender roles with a move towards women devoting more effort to domestic matters. This led to the growth of urban centres and the emergence of competing city states that ultimately gave way to the Akkadian and Babylonian empires. The first step in this process was the emergence of city states during the Ubaid and Uruk periods.

6.4 EGYPT: A PARADIGM FOR STABILITY

The vital part played by river valleys in human prehistory has become evident in earlier chapters. During the migration of modern humans out of Africa, once we ventured away from the coastline, the significant departures were probably along major rivers. When climatic conditions across Eurasia eased after 58 kya the routes away from the shoreline of the Indian ocean were up the mighty rivers such as the Euphrates, Tigris and Indus. These riverine environments offered not only attractive living space but also the logical pathways to new lands. Similarly, having reached the shores of the Black Sea, the Danube was the obvious route into Europe. Thereafter, in climatic terms, riverside settlements were usually the most reliable places for humans to live, even though there were risks of flash floods.

The benefits of riverside sites became even more evident with the stability of the Holocene climate. So, while Mesopotamia saw the earliest flourishing of city states, it was the climatic conditions of the Nile Valley that were to become the envy of the ancient world. Herodotus refers to 'the gift of the Nile', while the Roman writer Tibullus grumbles about the benefits of the Nile to Egypt: 'because of you [*the Nile*] your land never pleads for showers nor does its parched grass pray to Jupiter the Rain-giver.'

Although, as we will see, Pharaonic Egypt was hit hard by some of the climatic upheavals that beset other parts of the ancient world, of all the ancient civilisations it shows most clearty the benefits of the climatic stability that is the hallmark of the Holocene. There were times when serious drought led to damaging puctuations in the annual flood of the Nile from year to year and on longer timescales. Nevertheless, because of the immense scale of the watershed of the Nile covering both the Ethiopian highlands and much of equatorial Africa, as a general rule the interannual fluctuations of global and regional weather patterns were ironed out. So, in comparison to the climatic variations afflicting other regions that have been cradles of ancient civilisations, Egypt remained a region of unparalleled stability. When we examine the shift from the unreliability of the ice-age climate in mid-latitudes of the northern hemisphere to conditions during the Holocene, Egypt represents almost the complete opposite to the chaotic climate conditions before around 10 kya.

The stability of ancient Egypt raises interesting questions about the role of climate change in maintaining ordered social structures. In particular, the part played by religious beliefs in establishing a dominant theocracy seems to be an essential part of this stabilising process. The evidence of southern Mesopotamia suggested that increasingly elaborate rituals developed as part of maintaining complex agricultural systems. The relative stability of the climatic conditions that enabled Egypt to sustain a great civilisation for so long is mirrored in the permanence and rigidity of the religious structure that governed the country for millennia. There is a virtually unbroken thread in the unchanging symbols of religious authority extending over 2 500 years from the Old Kingdom to the Ptolemaic Pharaohs. In effect the country was ruled by an autocratic and unbending theocracy. Central to this system was the goddess Ma'at, who appears to have been the personification of the central concept of balance, order and truth that was the antithesis of chaos, and that guided the entire society.

The success of the religious structure must lie in part in the predictability of the annual flood. As noted in the previous section, the need to have extensive social structures needed to manage irrigation, are likely to have developed in Egypt, as in Mesopotamia. These arrangements are also likely to have taken on the same ritualistic form, even though the flood was more predictable. Indeed, once the leaders of the society had established a track record of success, their claims to possess supernatural powers and to be capable of interceding with the gods seemed all the more credible. At the same time the variations from year to year were sufficient to justify the priesthood's claims of the need for them to be funded to intercede with the gods on behalf of the people. Given that the chances of reasonable outcome were, by most standards, pretty high, the authority of the theocracy was reinforced over time. Only during periods of major disruption, such the first intermediate period between the Old and Middle Kingdoms, was the breakdown sufficient to destroy the power of both the theocracy and central pharaonic control structure, albeit temporarily.

If all this seems a bit far-fetched, there is considerable evidence of the dire consequences of the theocracy losing control at time of climatic disaster. There are frequent references to the collapse of religious principles during times of breakdown and stories of assaults on the priesthood. For example, the drought that appears to have destroyed the Maya civilisation in the eighth century AD (see Section 6.9) has been linked with gruesome evidence of the slaughter of priests and their families at the time. This savage retribution was presumably because of the failure of their rituals to bring relief from the parching conditions. Even as late as the nineteenth century a desperate community used the threat of violence. The *New York Tribune* carried the story of the citizens of a drought-stricken Mexican town who issued an edict saying that if rain did not arrive within eight days they would not go to mass or say prayers; if after another eight days adequate rain had not fallen they would tear down the churches and chapels and destroy other religious objects; and if relief had not arrived after a further eight days all priests, friars, nuns and saints would be beheaded. Fortunately for the Church heavy rain fell within four days.

Compare this situation with the shamanism that may have underlain the imagery in the cave paintings and mobiliary art the ice-age hunters. Whatever the overall purpose of these images, it is hard to imagine that there was not some element of seeking to influence the behaviour of the creatures portrayed. In a world where it was a matter of life and death to ensure that hunting parties intercepted migrating herds of large animals, any such potential influence must have seemed worth the effort. As discussed in Section 4.8, the ability to interpret the migrating patterns of birds and animals, together with the emergence of vegetation from winter dormancy, provided vital clues for making decisions. Dramatic changes in the weather from year to year exerted a profound influence on the timing of the return to summer breeding grounds. In the chaotic world of the LGM, attempts to invoke or influence the spirits of the animals to increase the success of hunting were immeasurably more risky than the equivalent activities of the Egyptian priesthood. So it is likely that the role of the shaman reflected a spiritual embodiment of the wishes of the community rather than exerting any control over group actions.

It follows from these observations that the potential for changing climatic conditions to influence the development of the religious practices is considerable. When viewed alongside the emergence of agriculture and the development of larger settlements promated by the climatic ameliovation in the Holocene, it seems that all these processes evolved together. So the sea change in climatic variability and the emergence of agriculture provide the ideal environment for human Spirituality to move from attempts to influence events by entering into the spiritual world to using such spiritual activities to establish agreed patterns of social control.

6.5 THE PRICE OF SETTLING DOWN

Outwardly, the experience of Egypt and Mesopotamia presents an unalloyed picture of progress and the benefits of agriculture. The emergence of climatic stability and more rigid social structures enabled people to live in larger communities and for many of them to develop specialist skills and trades. There was, however, a high price to pay in the terms of public health. The narrower diet that resulted from relying principally on a few cultivated crops (e.g. wheat, rice, maize, potatoes and yams) resulted in declining health. Archaeological evidence from around the world shows that, wherever populations became dependent on a staple crop and a diet that consisted largely of carbohydrate, health declined. The reason for this decline was the rising incidence of diseases related to shortages of minerals and vitamins (such as rickets due to lack of vitamin D and anaemia resulting from deficiency of iron).

Another health hazard in larger settlements was that living closer together increased the transmission of infections. More importantly, the close proximity of domesticated animals led to many diseases being transmitted to humans (e.g. influenza, smallpox, measles and mumps). It is estimated that over 300 diseases that now affect humans came from domesticated animals, of which roughly half came from dogs, cattle, sheep and goats (Karlen, 1995, p. 39). A considerable number of other diseases have crossed from wild animals: a risk that was also confronted by hunter-gatherers in their daily lives.

The consequences of this decline in health were manifold. The average height of adults declined significantly. Palaeopathological studies of skeletons found on the eastern Mediterranean region (Angel 1984) provide a measure of these changes. During the late Palaeolithic (30 to 12 kya) the average height of mature males found in sites from the Ukraine and the Balkans to Israel and North Africa was about 177 cm, whilst for women the figure was 166.5 cm. Even by the late Mesolithic (10 kya), these figures had fallen by about 5 cm. This decline increased during the late Neolithic (5 kya) to between 10 and 15 cm. Matters improved during classical times (500 BC to AD 500) when men had an average height of around 170 cm, and women 157 cm. These figures remained typical for northern Europe until the Middle Ages. There was then a decline in stature, especially in urban areas, so that in the industrial towns of nineteenth-century Europe grown men were on average 12 to 15 cm shorter than the modern humans who first occupied the continent some 35 kya. Part of this decline may have been related to the fact that many people did not need the same physique for specialised work in towns as for hunting or manual labour in the fields: a scribe or a tailor required dexterity and intellectual skills, not a strong back.

These broad figures may disguise an initial demographic improvement with agriculture. A recent study of the transition from Natufian hunting-gathering way of life to a Neolithic agricultural economy in the southern Levant, which examined 217 Natufian (10.5–8.3 kya) skeletons and 262 Neolithic (8.3–5.5 kya) skeletons (Eshed *et al.*, 2003), showed no indication of increased mortality with the advent of agriculture. On the contrary, life expectancy increased slightly from the Natufian to the Neolithic period. The transition to agriculture did, however, affect males and females differently: the mean age at death in the Natufian was higher for adult females than for adult males, while in the Neolithic, it was the reverse. One interpretation of this shift is that, with the onset of the Neolithic, increased fertility led to higher maternal mortality.

Although agriculture may have brought initial benefits, in the long run the effects of rising populations and an increasingly narrow diet had an adverse impact.Overall it is reckoned that since people started to live in settlements, reduced diet and infections killed more people than did war and famine. This estimate does, however, need to be treated with caution, as many examples of pestilence appear to have coincided with times of dearth. Where a population was weakened by famine, usually as a consequence of adverse weather, diseases were more likely to assume epidemic proportions. An even more interesting possibility is that periods of particularly adverse weather were linked to the emergence of new pandemics (e.g. the Justinian plague in AD 542 and the Black Death in AD 1348; see Section 6.9). It may be that a return to a more variable climate would increase the chances of such pandemics occurring, although it is to be hoped that modern medicine would be more effective in reducing their spread.

There was one benefit to the close physical proximity of man and animal. It eventually gave agriculturists a high resistance to animal-related germs. This resistance was a boon to agricultural communities that relied on animals for their livelihoods. It was, however, an unmitigated disaster for people living in parts of the world that were colonized by Europeans from the sixteenth century onwards. The diseases these colonists brought with them wreaked havoc with the native populations.

6.6 THE FIRST GREAT 'DARK AGE'

Although the Holocene has all the hallmarks of climatic stability, as has already been discussed in Chapters 2 and 5, there have been a number of notable hiccups in these calmer times. The idea that these climatic disturbances may have coincided with the occurrence of 'dark ages', when civilisations went into decline, has only emerged slowly. These dark ages were first identified clearly in the 1940s by the eminent French archaeologist Claude Schaeffer, who carried out major excavations of Enkomi, in Cyprus, Malatya, on the Turkish coast, and Ugarit, on the Mediterranean coast of northern Syria. He discovered various major destruction layers, which showed traces of fire and destruction that he originally identified as being caused by earthquakes. He then linked this work with the fact that many other places spanning a territory with a diameter of nearly 5 000 km (from Troy in the west to Tepe Hissar in the east, and from the Black Sea in the north to Lachish in the south) had been repeatedly destroyed during their Bronze Age existence. Up to four successive destruction levels were present in all sites, the most prominent of which were detected at the end of the Early Bronze Age ($\sim$ 4.3 kya), at the end of the Middle Bronze Age ($\sim$ 3.65 kya), and at the end of the Late Bronze Age ($\sim$ 3.2 kya). The first and the last of these three events coincide with the periods of rapid climate change identified in Section 6.1.

The extent of these events was far too great to be attributed to tectonic activity. A more likely explanation is that the climate change around 4.2 kya caused agricultural disasters and prolonged droughts that consequently led to a breakdown of social order in population centres in Asia, North Africa and eastern Europe. Furthermore, similar ecological and social upheavals appeared to have occurred at around the same time in China and the Americas.

Explanations of these periods of climatic deterioration tend to concentrate on either natural climatic variability or more specifically on large volcanic eruptions. Natural climatic variability has been attributed to an approximately 1500-year cycle in the climate of the North Atlantic, possibly of solar origin (Bond *et al.*, 2001). This is marked by notable cooling events in the North Atlantic, which appear most clearly in evidence of ice-rafted debris in ocean sediments in the vicinity of Iceland and Greenland some 4.2 to 4.5 kya, around 2.8 kya and 0.6 kya. The first of these has been linked with the events in the Middle East, and could be a factor in the sharp decline in rainfall around this time.

Evidence of volcanic activty is found in ice cores in Greenland and Antarctica. These show only three layers of acid tephra from major volcanic eruptions in this period, at 2354 BC, 1627 BC and 1159 BC. In addition, tree rings also show evidence of worsering climate at the time of these events. The event in 2354 BC has been attributed to the eruption of Hekla in Iceland. The event in 1627 BC has often been attributed to the eruption of Santorini, in the eastern Mediterranean, but analysis of tephra in the Greenland ice cores appears to rule out this connection. The 1159 BC event shows up in an Irish oak series (Baillie, 1995; Baillie, 1999) and also the tree-ring series in Anatolia (Manning *et al.*, 2001; 2003).

The most frequently cited of these upheavals is the rapid demise of Akkadian civilization (in what today is Syria) around 4.2 kya is associated with the onset of an unprecedented dry episode (Weiss *et al.*, 1993; 2001; deMenocal, 2001). Only a hundred years before the collapse, Sargon of Akkad had conquered the plains of Mesopotamia, and taken control of the Sumerian city-states, to establish a domain from the Persian Gulf to the headwaters of the Euphrates River. This was the first example of one state conquering other independent societies to form a single entity. When it was done the Akkadian empire controlled a substantial economic system. Trade extended from the silver mines of Anatolia to the lapis lazuli mines in Afghanistan, from the cedar forests of Lebanon to the Gulf of Oman. The empire's breadbasket was in northern Mesopotamia. A string of fortresses was built to control imperial wheat production. To the south, irrigation canals were extended, a new bureaucracy established and palaces and temples built from imperial taxes.

This empire lasted less than a hundred years. Evidence from Tell Leilan, in northern Mesopotamia, shows that the site was abandoned suddenly only decades after the city's massive walls had been constructed, its religious quarter renovated and its grain production reorganised. At Tell Leilan, analysis of the various layers of debris shows that after the Akkadian occupation there is an interval devoid of signs of human activity, containing only the clay of deteriorating bricks. The abandonment began about 4.2 kya. Soil samples from that time showed abundant fine, wind-blown dust, few signs of earthworm activity and much reduced rainfall. All this suggested that the people of Tell Leilan abandoned the site in the face of a sudden shift to a much drier and windier climate. This precipitated the collapse of the Akkadian empire's northern provinces. Only when wetter conditions returned, some 300 years later, was Tell Leilan reoccupied. This arid period is clearly seen as a dust increase in marine sediment cores from the Gulf of Oman. Using carbon-dating, the record shows that at 4.0 ± 0.1 kya, the abundance of dust jumped to two to six times above background, reaching levels not found at any other time in the past 10 kyr. The extreme dustiness covered a period of a few hundred years. Analysis of the isotopic composition of this dust pulse shows it to be similar to that of the soils of Mesopotamia and Arabia.

The onset of drier conditions around this time also brought famine to Egypt. A relief from the Unas Causeway at the Saqquara Pyramid shows the emaciated victims of famine. Later, from the tomb of nomarch Ankhtifi, near Luxor, there is a quotation about one of the earliest-known great famines (around 4.15 kya):

> '... All Upper Egypt was dying of hunger, to such a degree that everyone had come to eating his children, but I managed that no one died of hunger in this nome [province]. I made a loan of grain to Upper Egypt ... I kept alive the house of Elephantine during these years ...'

Although Egypt was protected from the worst climatic calamities by its unique position (see Section 6.4), it evidently suffered grievously during this time. The decline in the floods appears to have been accompanied by a shift in weather patterns with hot winds from the south bringing dust storms that obscured the sun so that men could not see and '... none may know that it is midday, and the sun will cast no shadow.' This disruption led to a breakdown in the economic system in Egypt. There was a plethora of pharaohs during the three decades of the Seventh and Eighth Dynasties, each ruling for a year or two and disappearing without trace. In these conditions, 'when the Nile was empty and men crossed over it on foot', Egypt splintered into a number of feudal states.

In the period of breakdown local governors appear to have taken responsibility for their own provinces, by conserving water supplies, and reducing the number of hungry mouths by repelling famine-stricken outsiders: Libyans, and the Bedouin of Sinai and the Negev, who grazed their flocks on the borders of the delta, in the manner of Abraham and Jacob, when there was famine in their own lands. It was a desperate struggle. In the worst years, when the floods failed and food supplies dwindled to virtually nothing, cannibalism was rife. Monuments of the period, which are and far between, are crude tomb stele that record the boasts of local rules who presided over these terrible times.

The evils caused by famine, poverty, social upheaval and anarchy brought others in their train such as plague and a drop in the birth rate. A deep and lasting impression was left on the ancient Egyptians by the trauma of these times, so that in later literary works, such as the Prophecy of Neferti and the Admonitions of Ipuwer,[3] when the writers wished to depict mankind tormented by intolerable miseries, it was the sufferings of this period that they recalled.

As noted above, the evidence of rapid climate change around 4.2 kya appears to be restricted to southwest Asia and Egypt. In addition, it is almost certainly linked to the permanent shift to drier conditions observed in northern Nigeria (Street-Perrott *et al.*, 2000) (see Section 5.17) that occurred around 4.1 kya. Although there are some claims that the civilisation on Crete and mainland Greece went into decline around this time, the evidence is limited and lacks support from local climatological proxy records. Similarly, other records

[3] The Prophecy of Neferti is set in the Fourth Dynasty but may have been written much later and purports to predict Egypt's decline and the coming of a saviour. The Admonitions of Ipuwer appear to have been written around the end of the Middle Kingdom and the Second Intermediate period and to recount the privations of the First Intermediate Period. Both works are generally believed to have served as a reminder to Egypt of the disorder that follows in the train of famine.

do not substantiate claims that the great cities of Mohenjo-daro and Harappa in the Indus Valley collapsed around the same time. But, as described in Section 5.3, the drying of lakes in the Thar Desert (Enzel *et al.*, 1999) preceded by nearly a thousand years the rise of the Early and Mature Harappan phases of the Indus civilisation from 4.6 kya to 4 kya. Improved climatic conditions did not lead to the rise of this major urban civilisation. Furthermore, the collapse of the Indus culture around 3.7 to 3.9 kya did not coincide with a more arid climate. The Thar data indicate that the Indus civilisations flourished mainly along rivers during times when northwestern India experienced semi-arid climatic conditions that are similar to those at present.

6.7 THE DEMONISATION OF THE PIG

The impact of climate change on the social structures was not just a matter of the rise and fall of ancient civilisations, and their associated social structures. There is evidence that it took on more subtle forms. The establishing of large-scale agriculture had a wide range of economic and social implications. The emergence of food taboos in societies of southwest Asia around 5 to 4 kya may be an interesting example. This is yet another area of intense academic speculation and controversy, where an examination of the impact of climate change may shed some new light.

Nowhere is the question of food taboos more fiercely debated than in respect of the humble pig. In many temperate parts of the world the pig is revered as an excellent means of dealing with a number of agricultural challenges. They are good at living off mixed forest and poorly drained land where they thrive off tubers, roots, nuts and fruits. They cannot, however, make effective use of high-cellulose material such as husks and stalks. In hotter drier climates, where shade is at a premium, they are less efficient and need their diet supplemented by grain, which might otherwise feed humans. Worse still they suffer from sunburn and hence need plenty of shade.

All these factors suggest that the pig may not be the ideal domesticate for the hot dry climates of the Middle East and northern

India (Wenke, 1999, pp. 280–1). Furthermore, pigs cannot be milked, sheared, ridden or used as draught animals. Nevertheless, many anthropologists do not accept these economic arguments. Instead, the traditional explanation is that the religious proscription of the pig was a matter of public health. Because the pig is often infected by trichinae, parasites that can kill people, this is cited as the reason why pigs were banned. The fact that cattle, sheep and goats all carry diseases that are far more likely to infect humans does not seem to intrude into this thesis.

An alternative explanation is that it was a climatic factor that sealed the fate of the pig in the Middle East. This is the upheaval that occurred in the late third millennium BC. There is widespread evidence of the extensive domestication of the pig in southwest Asia from 8 kya onwards. As late as 4.9 kya pig bones represented 20 to 30% of all mammal remains in many large sites. Then, sometime after 4.4 kya, pork apparently became religiously proscribed in most Mesopotamian cities and in Egypt. The fact that the timing of this change coincides with the hotter drier climate that led to the collapse of the Akkadian Empire may explain the elimination of poor piggy. At a time of heat and drought, the failure of the pig to deliver, and worse still its competition for valuable resources that fed humans, provided the last straw. So, while the purely economic argument may not be adequate to explain this major shift in agricultural practice, when combined with the disaster of prolonged drought this may have made it much easier to attach a religious reason to the demonisation of the pig.

6.8 THE SEA PEOPLES

At the end of the late Bronze Age (around 3.3 kya/1300 BC) the economy of the eastern Mediterranean was thriving. Evidence of the scale of this activity is found in the extraordinary cargo of a wrecked ship found at Uluburun near Kas in southwest Turkey (Pulak, 1998). On the basis of tree-ring analysis, the ship is dated as having sunk around 1306 BC. The cargo comprised mostly raw materials, the most extensive of which was approximately 10 tons of copper ingots. There was also a ton of terebinth resin, possibly used as incense, and 175 glass ingots of cobalt blue, turquoise and a unique lavender colour. In addition, there were logs of Egyptian ebony, ostrich eggshells, elephant tusks, hippopotamus teeth, opercula from murex seashells (another possible ingredient for incense) and modified tortoise carapaces (almost certainly sound-boxes for stringed musical instruments). Finished products found at the site have been linked to at least nine cultures from the Baltic to Nubia, and include Cypriot pottery, metal vessels, wooden containers, beads, and cloth or garments, tools of trade (weights), possessions of the traders (e.g. swords and seals) and jewellery including a gold scarab bearing the cartouche of the Egyptian queen Nefertiti.

This treasure trove is testament to the scale of economic activity at the end of the Bronze Age. It also lends support to the thesis that the Trojan War is a memory of the conflict between the armies of Mycenae and the Hittites, which fought over Troy, as the city would have exerted control of trade in and out of the Black Sea. This booming economy ground to a halt in a period of globally disturbed climate. The climatic upheaval, which appears to have had widespread ramifications for ancient civilisations, occurred between 3.5 and 2.5 kya. Across Europe colder wetter conditions set in around 3.4 kya and the glaciers in the Alps started to expand. In northern Scotland there is a distinct and large-scale shift to wetter climatic conditions. The transition appears to have occurred abruptly, possibly over a decadal to century timescale. Broad correlation with deep-sea sediment records suggests that the transition may reflect colder sea surface temperatures in the North Atlantic. At the same time the eastern Mediterranean cooled markedly, initially in the form of colder winters, which probably led to an increasing incidence of drought in the region, and ushered in the next Dark Age.

These changes appear to have triggered large-scale demographic movements that are usually associated with the name 'Sea Peoples', in spite of the fact that their major advances were overland. They are best known for waging two campaigns against Egypt, which ended in battles in 1232 BC and 1183 BC. Although they were repulsed on both occasions, after these attacks Egypt ceased to be an imperial power. Other civilisations were even worse hit. The Hittite kingdom in central Anatolia, which had represented a serious challenge to Egypt during the preceding century, was totally extinguished in its Anatolian heartland, although there was some continuation of the culture in Syria. In Greece the might of Mycenae vanished.

The Sea People were probably part of a great migration of displaced people. The migration may well have been the result of widespread crop failures and famine relating to the climatic change occurring at the time. They were, however, an efficient military force: aggressive, well-armed and ruthless raiders. Their successful progress appears to have focused on attacking capitals and cities important to administration. In these cities they destroyed government buildings, palaces and temples, while leaving residential areas and the surrounding countryside untouched. They appear to have first destroyed Mycenae, and then moved on to Troy, which they laid waste around 1250 BC. They then moved into the Levant and on to Egypt where they met their match in the two battles mentioned above.

The Sea People were always presented as a negative and destructive force for the region. This may be unfair. True, the Sea People destroyed much through their campaigns. On the other hand, they were probably the founders of the Philistine and Phoenician civilizations, which grew to be included among the most important forces in the eastern Mediterranean when the region emerged from the dark age at the beginning of the first millennium BC. In any case, if they were driven by climatic events, it is hardly accurate to attribute their behaviour to implicit bellicosity. Whatever the driving forces, it was centuries before the economic activities in the region returned to a level that matched the eclectic nature of the cargo found at Uluburun.

6.9 THE CONTINUING CATALOGUE OF 'DARK AGES'

In spite of the arbitrary chronological cut-off at around 2.5 kya for detailed analysis of past events, it is worth noting briefly that the collapse of civilisations did not end at this point. Indeed the periods of rapid climate change identified in the introduction to this chapter may well have played their part in history. The Dark Age of the seventh century AD in the Byzantine empire is a good example. Here climate may be but a small part of a multifaceted disaster. Although there were signs of social stress in the Byzantine empire in the early sixth century, possibly as a consequence of population pressures, the arrival of bubonic plague in AD 542 may have killed as much as a third to half of the population. It then returned with terrifying regularity, every 15 years or so, to most of the major cities. By the end of the sixth century the scale of depopulation was profound, with many cities that had survived since antiquity ceasing to exist. By the mid-eighth century the population of Constantinople had sunk to between 25 000 and 50 000 from a figure of some ten times this at the beginning of the sixth century. Economic activity throughout much of the Mediterranean world virtually ceased and the region slipped back into a form of rural convalescence, with life continuing more easily in the countryside where contagious diseases exerted a less deadly sway.

The climatic connection is that a more cataclysmic event may have triggered this decline. There is widespread evidence of the Sun being obscured by a 'mystery cloud' in AD 536. It is usually assumed this was caused by a massive volcanic eruption (Stothers, 1984), possibly of Rabaul in New Guinea, that, in the same way as Toba some 70 kyr earlier, created a global dust veil in the upper atmosphere, which absorbed sunlight at high levels and cooled the Earth's surface. Chroniclers from Rome to China record that the Sun dimmed dramatically for up to 18 months and there were widespread crop failures. There is plentiful evidence of a significant deterioration in the climate for several years in tree-ring data, but the evidence of a truly massive volcano in the Greenland and Antarctic ice cores is not convincing (Clausen *et al.*, 1997). The alternative explanation of the mystery cloud being caused by a comet is the subject of much debate, but is not supported by convincing physical evidence of its impact (Baillie, 1999).

Another sudden collapse is that of the classical Mayan culture. This is one of the great mysteries of prehistoric Mesoamerica. Having emerged around 1500 BC this civilisation thrived from around AD 250. It is renowned for its monumental constructions and reached a pinnacle in the eighth century. By this time the population density in the Mayan lowlands (which extend over modern-day Guatemala, Belize, Honduras and Mexico) was far higher than current levels. It sustained a sophisticated society that built magnificent buildings and other edifices. But early in the ninth century the civilisation entered a cataclysmic period. The monumental construction and detailed records came to an abrupt end at one centre after another. This decline tallies closely with evidence of extreme drought in the region that comes from analysis of local lake deposits and from ocean sediment cores from the Cariaco Basin off the coast of Venzuela (Hodell, Curtis & Brenner, 1995; Haug *et al.* 2003). This period of drought, which appears to have been the most severe in the last 7 kyr, also coincides with the peak in ENSO activity in the same period (see Section 5.19).

There are various examples of more recent climatic fluctuations, notably during the Little Ice Age (see Section 6.1), which led to subsistence crises in Europe. As a general observation, these were of less extreme nature than the dark ages that have already been cited. This can be regarded as evidence that our society has become better equipped to handle the levels of climatic variability typical of the Holocene. Alternatively, it could be that we have been lucky in the last millennium or so. This is a question we will return to later.

7 Our climatic inheritance

> At the base of all these aristocratic races the predator is not to be mistaken, the splendorous blond beast, avidly impatient for plunder and victory.
>
> Friedrich Nietzsche (1844–1900), *Zur Genealogie der Moral*

Nietzsche's observation about the superiority of certain races is an anathema to modern political thinking on racial equality. This repudiation is not just a matter of political correctness, but reflects the fact that his ideas are wholly contrary to the evidence of genetic mapping presented in earlier chapters that shows how we are all so closely linked in terms of genetic origins. As Svante Pääbo, of the Max Planck Institute for Evolutionary Anthropology in Leipzig, Germany, recently observed in a review article (Pääbo, 2003), '[evidence] suggests that we expanded from a rather small African population. Thus, from a genomic perspective, we are all Africans, either living in Africa or in quite recent exile from Africa.' Nevertheless, in exploring how modern humans and their social structures evolved from the depths of the ice age and how climate change may have left an indelible imprint on our make-up, we will need to consider the more violent aspects of human behaviour. So, while we can have no truck with Nietzsche's notions of 'aristocratic races', blond or otherwise, the issues of the 'predator' and of 'victory and plunder' have been a major feature of the analysis our prehistory and more recent history.

The objective of this chapter will be to explore the implications for human development of the characteristics of ice-age people (i.e. some combination of powerful physique, stamina, intellectual development and social organisation) being markedly different to the requirements of subsequent societies. We need to ask a series of questions about the extent to which these pressures, as a result of natural selection, have resulted in a permanent change in our make-up. In a world where some of the restraints on what might be regarded as more primeval behaviour are shifting, it is important to understand the origin of these fundamental evolutionary features of our character and how they moulded many features of our society. Central to this analysis is appreciating how the type of world out of which we evolved is imprinted in us, and can powerfully influence how we respond to opportunities and challenges.

7.1 DID WE HAVE ANY CHOICE?

In asking basic questions about the impact of climate change on the evolution of our social structures, apart from eschewing any tendency to adopt a racial position, there are two other matters that require careful handling. The first is what might be termed the 'Paradise Lost' interpretation of the world we left behind. This is something that we subsequently utterly destroyed for many of those hunter-gatherer societies, which had survived until recently in the less-populated parts of the world. The sense that the transition from hunter-gatherer society to agriculture and then industrialisation led to a decline in our quality of life was first encapsulated by Marshall Sahlins, Professor of Anthropology at the University of Chicago, in a chapter entitled 'The Original Affluent Society', in his book *Stone Age Economics*, (Sahlins, 1972). This negative view was also taken up by Jared Diamond (Diamond, 1987) in a paper entitled 'The worst mistake in the history of the human race.' He concludes by asking the question: '... will the plight of famine-stricken peasants gradually spread to engulf us all? Or will we somehow achieve those seductive blessings that we imagine behind agriculture's glittering façade, and that have so far eluded us?'

The opposite view has been rather pithily expressed by Dan Usher, Emeritus Professor of Economics at Queens University, Kingston, Ontario, in the first chapter of his book *Introduction to Political Economy* (Usher, 2003), entitled 'How Dreadful Life Used to Be'. As he observed, 'To be constantly on the move by foot with no vehicles whatsoever, to be without solid housing, with bonfires as the only mode of heating and lighting, without writing, radio, television or electricity, without clothing other than what one makes for oneself, without medical care, and entirely ignorant of the world beyond one's field of vision, is a life that only an anthropologist could admire.' The debate about the benefits of the hunter-gatherer lifestyle in the 1970s and 1980s was part of a wider question about whether food production would keep pace with rapidly rising population levels. Many environmentalists assumed that it would not and that this would lead to a global demographic crisis. More recently, this concern has tended to be subsumed in the wider concerns about global warming and the general destruction of the environment.

The message that does emerge from the nature of the climatic changes at the end of the ice age is that the combination of the events surrounding the Younger Dryas, allied with the subsequent fundamental decline in variability, exerted a powerful influence on social developments. The onset of a warmer wetter climate during the Bølling and Allerød provided opportunities for increased sedentism among hunter-gatherers in the Middle East where there were adequate year-round supplies of food. With this tendency to settle down came a rise in population. The Younger Dryas then provided an immense challenge to this emerging lifestyle that seems to have stimulated the first steps towards agriculture. The return to a warmer wetter climate at the end of the Younger Dryas together with a dramatic reduction in variability from year to year at the start of the Holocene made agriculture overwhelmingly attractive. It wasn't a matter of 'making a mistake', it was *force majeur*: in effect there was no alternative.

The spread of agriculture across Europe during the first half of the Holocene and the rapidity with which it was taken on board supports this fundamental conclusion. The alacrity with which people switched to a diet of domesticated plants with the arrival of agriculture in northern Europe is striking. For example, in Britain there is evidence of a rapid and complete shift from marine- to terrestrial-based diet among both coastal and inland dwellers at the onset of the Neolithic around 6 kya (Richards, Schulting & Hedges, 2003). This suggests a sudden adoption of agriculture and animal husbandry rather than a gradual move to the new way of life. It implies that the new farming lifestyle must have been sufficiently attractive to persuade even coastal communities to abandon their successful fishing practices.

All this supports the sense of inevitability about the switch to agriculture. Although the long-term implications of the shift may have been a Faustian bargain, at the time the immediate attractions were compelling (Olsson, 2001). It provided an escape from a Malthusian trap, albeit temporarily, permitting unrestrained growth in populations. Once established in particularly well-endowed locations, it exerted irresistible pressures on the hunter-gatherer activities by driving both social structures and new technologies in the direction of agriculture. Only when the new societies started to bump up against a new Malthusian ceiling did the economic disadvantages of agriculture start to emerge. But, by then, it was too late to change, and in any case, as Maynard Keynes observed: 'In the long-term we are all dead.'

Lurking behind these basic observations about population change is the more complex issue of the cultural consequences of the move to agriculture. A striking feature of our current social order is the extraordinarily widespread and enduring nature of our cultural diversity. This is in stark contrast to the comparatively small variation in our genetic make up, which has been discussed earlier. Extreme cultural diversity appears to be the product of our hunter-gatherer past. Although this diversity may have shrunk appreciably as a consequence of burgeoning agricultural communities, it remains amazingly varied (Pagel & Mace, 2004). As such it probably remains an enduring relic of our ice-age past. But how it survived the transition from hunter-gatherers to farmers falls outside the remit of this book.

A second pitfall to avoid is seeking to establish climatic connections to the more fundamental political and philosophical aspects of the development of the societies discussed in this book. Earlier discussions of the human price paid for the development of agriculture could easily lead us on into Jean-Jacques Rousseau's theories as expressed in the Social Contract and its famous opening words that 'Men are born free, yet everywhere they are in chains' or into exploring the concept of the 'noble savage'. Equally, it would be unwise conversely to embark on a philosophical route that sought to demonstrate that grappling with the challenges of climatic variability engendered creativity that was to lead in time to Michelangelo, Mozart or Einstein. I have neither the space nor the competence to do justice to these issues. All that can be done here is to examine whether there are latent features in our physique, intellectual development and psyche that are the product of our ice-age evolution: challenge enough without getting into more dangerous philosophical quicksands.

7.2 REGAINING OUR PALAEOLITHIC POTENTIAL

It is a standard view of our current physical condition that we assume that we are much bigger than our forebears. The evidence of medieval armour is often cited as showing that even the aristocracy in the Middle Ages were diminutive compared with present-day men. Indeed it appears that, based on European figures, our stature fell to its lowest levels in the seventeenth and eighteenth centuries and has been rising ever since. As discussed at length in both Chapters 4 and 6 the decline in human size appears to have been steady from the Early Upper Palaeolithic onwards. This extended into the Neolithic and is seen as part of price of agriculture and settlement because the average quality of nutrition declined and the incidence of disease increased: living close together and in the company of animals was not as healthy as the hunter-gatherer lifestyle (Karlen, 1995, p. 35).

Even more striking was the stark fact that being poor was not good for one's health. By the start of the First World War the recruits to the ranks in the British Army were on average 15 cm shorter than their officers who came from the wealthier classes. When losses rose, the recruitment criteria were reduced and the army formed 'bantam regiments' in which every man was between five feet and five feet three inches tall (152 to 160 cm). In fact, many of these regiments were recruited from mining communities, and there was considerable evidence of there being benefits in being small when working down the mines (Kirkby, 1995). In these communities, stature may not have been a true measure of physical deprivation, but more a case of smallness and hardiness being a benefit. Nevertheless, that so many tiny men could be recruited is still a telling commentary on the physical condition of working people in Britain at the beginning of the twentieth century. Furthermore, it does not alter the fact that these plucky little men had to carry nearly 40 kg of equipment as they charged through the mud of the Somme and Passchendale into a withering hail of machine-gun fire.

In a world that had come to regard the progress of the nineteenth century as a natural product of the 'ascent of man', these examples do not sit easily with the figures quoted in Chapter 4. In the Upper Palaeolithic those modern human males who survived to adulthood were on average big chaps. What is more they were a step down from their cousins, the Neanderthals. Modern rugby players would probably have blanched at having to pack down against the average mature Neanderthal, whose combination of physique and weight would have made him ideally suited for a heavyweight wrestler. What is equally clear is that in the developed world, during the second half of the twentieth century, the improvement of diet, notwithstanding the increasing availability of junk food, has led to a dramatic increase in the height of people, not to mention their girth. While the latter would not have been an asset to the people of the Palaeolithic, it is probable that their height and weight represents an indication of what might be regarded as normal for our species. The inference may be that the generations growing up at the beginning of the twenty-first century in the developed world will find that the increase in stature, which has been a feature of the preceding couple of generations, will slow down. In short, they will be achieving their palaeolithic potential, and hereafter, growth in size is more likely to be around the waist than upwards.

In case this all sounds a bit too much like a campaign for weightwatchers, there is an interesting underlying issue. This is the nature of our omnivorous past. As was clear in Chapter 3 our palaeolithic ancestors exploited a wide variety of food sources. This breadth of diet was in stark contrast to that adopted by many agricultural communities from the mid-Holocene onwards. The shift to relying on cereals and dairy products, plus the more recent shift to a greater reliance on meat, is often cited as the cause for many modern ills. As was noted in Section 4.5, the measurement of the stable isotopes of carbon and nitrogen in samples of human bone provides direct evidence of past diet (Richards *et al.*, 2001). Limited studies on five Neanderthals from three sites, as well as a number of Upper Palaeolithic and Mesolithic humans, confirms the importance of animal protein in diets, although there was a marked broadening of the diet after the end of the ice age.

There is one other interesting physiological consequence of the shift in diet associated with adoption of agriculture. This development has the rather ugly description of 'cranial gracilization' to describe the fact that we have become prettier. Our jaws have got smaller and our mouths have become more crowded as a result of chewing less hard food and preparing soft foods that have been cooked for lengthy periods. These changes have not led to serious health effects, but the reduced space for our teeth has increased the incidence of dental problems. At the same time our brain size has reduced by about 10 percent.

A much more provocative interpretation of the changes in brain size comes from Harvard University anthropologist Richard W. Wrangham. He argues that they are part and parcel of humans becoming more domesticated. As a result of humans becoming more sedentary and their populations growing, over the past 20 kyr, there were selection pressures for greater within-group peacefulness and within-group sexuality. Driven by the challenge of limited resources, there were obvious pressures that led to the reinforcement of within-group cooperation and between-group competition. This led to within-group amity and between-group enmity among humans. The most profound consequence of these pressures may have been the reduction in our brain size (see Section 4.4), but this selection process may also have contributed to our smaller jaws and teeth. At the same time, it may have led to our year-round breeding season and prodigious sexuality: the Holocene climate has a lot to answer for.

So far the discussion has been in terms of the general physiological implications of climate change on humankind. There is, however, the much more contentious issue of how, in response to both climate and diet, genetic factors may have led to both historical and geographical differences in physiology of different cultural groups. Although this gets us into deeper water, it is an area that cannot be ignored. Two essential changes have taken place since modern humans migrated out of Africa. First, there is the pigmentation of our skins. Second, there are the physiological changes in body size and proportions that appear to reflect the efficient thermoregulatory response to different climates around the world.

In respect of skin pigmentation the genetic arguments are equivocal. Human hair and skin pigmentation is a highly visible trait and is the primary protection against the damaging effects of ultraviolet radiation (UVR). Little is known, however, about the genetic variation responsible for the large array of pigmentation observed in human populations (Rana *et al.*, 1999). Two classes of melanin, the red/yellow phaeomelanins and the black/brown eumelanins, are present in the epidermal layer of human skin and hair. The spectrum of human hair and skin pigmentation observed in different geographic regions of the world is the result of varied production, distribution and packaging of these two classes of melanin. The type of melanin that is produced in our bodies depends on both genetic factors and the amount of UVR absorbed by the skin.

The most obvious benefit of high levels of melanin in the skin relates to sunburn. In sunny parts of the world dark skin was a vital factor in our evolutionary development. At higher latitudes the presence of high levels of melanin can be a disadvantage, notably in reducing vitamin D production. Genetic mapping studies provide some support for the theory that skin pigmentation variation observed today is related to the production of this vitamin in the body. Vitamin D is produced through conversion of the dietary precursor, 7-dehydrocholesterol, by UVR, in the skin capillaries. Dark skin reduces the amount of UVR reaching the capillaries, which in turn reduces the amount of vitamin D formed. In regions of low sunlight, this can lead to vitamin D deficiency and the childhood disease of rickets, which in more severe forms can reduce survival and reproduction. For those living in the tropics, dark skin can be an advantage as excess vitamin D may be toxic. It is possible that African populations retained their ancestral sequence to maintain a darker pigmentation for protection against both UVR and the toxic effects of vitamin D. As populations migrated to regions of lower sun exposure, the selection pressure for dark pigmentation was relaxed and genetic mutations that result in lighter pigmentation might have increased in frequency.

Recent studies of one of the genes that may be important in the selection process governing pigmentation, Melanocortin-1 receptor (MC1R), have thrown some light on the evolution of skin colour (John *et al.*, 2003). MC1R has been implicated in the red hair and pale skin of human non-African populations. Mutations at MC1R that result in lighter pigmentation might be advantageous and increase in frequency among people living at higher latitudes. This may explain why there is a higher variation of MC1R in non-African populations. Conversely, a lower level of mutation in this gene has probably played a significant role in the maintenance of dark pigmentation in Africans. It would appear that genetic studies are beginning to unravel the mechanisms behind the evolution of skin pigmentation of people around the world.

As for the evolution of a metabolism to cope with the cold climate, more efficient energy metabolism and large body size are advantageous (see Section 4.4). In moving out of Africa and into colder climates, the likely selection process was to favour mutations that enabled people to generate more body heat. While less efficient in terms of using available energy, it may have been vital in surviving the cold. The consequence for the descendants of these people is that in centrally heated homes they are still programmed to generate excess body heat, so when this is not needed for comfort, it is stored for some future challenge in the form of fat. So our current concern with what is seen as an epidemic of obesity may be laid at the door of the genetic changes that occurred when we first took on the challenge of ice-age Eurasia.

Alongside the issue of how our physique has changed since palaeolithic times, there is the more specific question of just how much our diet has changed. The inference from the existing studies is that the breadth of the palaeolithic diet and its heavy reliance on animal protein was healthy, the assumption that the introduction of a cereal-based diet led to a general decline in health is much less easy to substantiate. The switch away from the broad plant-based content of the diet in the Upper Palaeolithic and Mesolithic may have been the important factor in these changes. Even so, any attempt to assert that the replacement of this broad plant-based diet by a much narrower cereal-based diet led to the declining health of the Neolithic peoples takes us into controversial territory. For example, the current debate about the merits or demerits of the high-protein, low-carbohydrate Atkins diet has embraced the issue of whether the evolutionary consequences of the palaeolithic diet may make some of us better suited to indulge in this particular fad. Proponents of the diet claim that as modern humans evolved during the ice age surviving on a high-protein, high-fat diet, this undermines the health arguments that such a diet is dangerous. On the contrary, they assert that it is the replacement of the palaeolithic diet by one high in refined carbohydrates that has done the damage.

7.3 WARFARE

After what was said in Chapter 4 about the climatic controls on gender roles in the Palaeolithic, it is tempting to look for a thread that leads from the male as a pursuer of big game to the endemic nature of warfare in human history (Haas, 2002). The emphasis on male characteristics of strength, speed, throwing ability and aggression could be seen as the obvious combination for continued belligerence, even when we settled down to life as farmers. In effect, is there any justification in attaching a climatic link to the human affliction that Hobbes described as a 'state of war by nature'? From the outset it has to be stated, however, that there is scarcely any evidence of warfare during the Palaeolithic. Maybe it was more a case of not having the time to develop this aspect of our character because of the demands of life in the ice age, which, to paraphrase Hobbes again, could be described as being 'solitary, poor, nasty, brutish and short'. Here we are in danger of being drawn into philosophical issues about human nature that are beyond the scope of this book. Hobbes' empiricist view is fundamentally different from Rousseau's 'romantic' view (see Section 7.1).

In exploring Palaeolithic warfare, the first point to make is that absence of evidence is not the same as evidence of absence. The lack of signs of warfare in the Palaeolithic may merely tell us that such activity at the time did not involve structures (e.g. defensive systems) that left lasting evidence. The case rests principally on examples of human remains that show signs of a violent death, of which there are precious few examples in the Palaeolithic. In addition, there is no significant evidence of conflict in the art of the Palaeolithic. All of this seems to point in the direction of Rousseau. The low population densities did not generate the pressures that lead to friction between groups. This suggests that only when we started to compete for resources did we develop a propensity for warfare.

The first extensive example of violent deaths in the archaeological record emerges well after the end of the LGM. This could be seen as evidence humans turned their efforts to warfare only when they had time on their hands. As life became that little bit more manageable and populations grew, they diverted energy that had previously been devoted to surviving, into stealing each other's belongings. This alternative view does not seem particularly likely, given how much spare time our ancestors living on the steppes of Asia evidently had for producing personal ornamentation, as the burial at Sunghir demonstrates (see Section 3.11). So an explanation based on competition for resources appears more likely.

In Europe there are only scattered examples of human remains showing a violent end during the period from 15 to 10 kya. On the other hand, in Egypt there is a cemetery – Jebel Sahaba – used by the local hunter-gatherers repeatedly from 14 to 12 kya that contains clear evidence of violent deaths (Wendorf, 1968). Nearly half the bodies excavated from this site show signs of violence (e.g. projectile points embedded in their bones, depressed fractures of the skull and fractures of the forearm consistent with parrying blows). The intriguing feature of this site is that, although there is no evidence of a local settlement, the repeated use suggests a territorial aspect to the lives of those who lived there at the time. It is this territorial aspect that may hold a key to the nature of early conflict. Perhaps, when the Nile valley became deeper and then much narrower with the torrential flow that followed in the train of the warmer wetter climate around 14 kya (see Sections 3.14 and 5.17), then competition for limited resources in the region grew to antisocial levels.

In contrast to this rather stark picture of Epipalaeolithic life in Egypt, extensive studies of the remains of the Natufian hunter-gatherers (see Section 5.8), who lived around 14.5 to 12 kya, show virtually no evidence of warfare. Analysis of several hundred skeletons shows that only two have any signs of trauma, and nothing to suggest that these injuries were the result of military action (Ferguson, 2003). This is in some ways surprising, as another feature of this culture was an increasing population density, which appears to have the been the source of conflict at Jebel Sahaba.

The preconditions to warfare appear to have been a rising population density, resource shortages and perhaps ethnic differences. This in turn led to competition over territory and possessions. The issues of squabbling over territory appear to become the core of conflict after around 10 kya in Mesolithic northern Europe. Analysis of burials at Skateholm in southern Sweden dating from around 7.5 kya shows a remarkably high level of violent deaths (Mithen, 2003, p.175). Here, and elsewhere in northern Europe at this time, over a fifth of skeletons show evidence of a violent end. The proportion is higher amongst men, and certain injuries, such damage to the left-hand side of the skull and arms broken parrying blows, suggest conflict. In the case of Skateholm, the site was rich in wildlife around 7 kya and was favoured as a winter refuge. At that time rising sea levels had flooded what is now the North Sea. The consequent loss of resources probably led to conflict within an expanding population over control of the best sites. This aggression may also have been an early example of conflict between the separate cultural identities of different groups.

The same story of pressure on resources emerges from analysis of rock art in northern Australia (Taçon & Chipperdale, 1994). This tells a tale of continuing collective violence, which develops from individual and small group conflicts around 10 kya to larger group confrontations from about 6 kya. This covers a period of ecological crisis when the rising sea flooded the rich plains between Australia and New Guinea. Here again the combination of reduced resources and possibly rising population levels appear to have been the initial catalyst for conflict.

The next stage in the emergence of conflict was the development of agriculture and the associated growth in population. Any territorial factors that might have emerged with the competition between hunter-gatherer groups for seasonal encampments would have been compounded by the fixed nature of agricultural communities. Even so, the incidence of warfare seems to ebb and flow through the archaeology of Neolithic societies. Moreover, even where examples of defensive structures start to emerge, the evidence has to be treated with great care. The interpretation of the dense interconnected structure of Çatalhöyük as a defence against attack is plausible (see Section 5.9). It is, however, salutary to observe that perhaps the best known of all early defensive structures – the walls of Jericho – may have originally been built as flood defences (Bar-Yosef, 1986). It was only when centralised, hierarchical societies began to emerge that the almost continual warfare that seems to have been such an integral part of recorded history became established.

Looking for a climatic component in warfare becomes much more difficult if we cannot show that this behavioural pattern was an integral part of surviving the ice age. Indeed, if anything, the evidence of the Upper Palaeolithic is that the gregarious nature of modern humans was an important feature of their survival in Europe at least. When it comes to more recent history, the climatic component appears more likely to have lain in adverse conditions driving human movements that led to conflict. Inevitably those fleeing from some natural disaster, be it rising sea levels, drought or famine, encroached upon the territory of other peoples. Conversely, as was noted in Chapter 6, Sargon of Akkadia's initiative of consolidating the city-states of Mesopotamia may, in part, have reflected the growing prosperity of the region and the attraction of acquiring the possessions of other peoples'. While this newfound wealth may have been attributable to benign climatic conditions, it is ironic that this first real empire-building enterprise based on the effective application of warfare should apparently be laid low by climatic change. This collapse did not, however, interrupt the movement of ancient civilisations to embrace warfare as the principal means of achieving their economic goals.

Nevertheless, resource constraints may have continued to drive major population movements. It is generally accepted that the migration of the Sea People (see Section 6.8) resulted from some form of collapse in their agricultural system, possibly as the result of adverse climatic developments in the home territories. While their objectives were economic, they were spurred on by necessity rather than greed.

This line of argument leads to a more tenuous connection between the skills that evolved as part of survival in the ice age and our subsequent propensity for warfare. This is that the technologies that we developed to hunt animals and the intellectual skills of hunting equipped us to be highly efficient killing machines. As a consequence when eventually we turned our mind to warfare, it is hardly surprising that we turned out to be rather good at it. But it was only when we were pressed together in larger numbers, and confronted with shortages, that this propensity was developed with such devastating efficiency.

On the other hand, the evidence of extensive palaeolithic social networks may constitute a compensating benefit. This structure would have required the development of social skills such as the formation of alliances and the exercise of diplomacy to defuse potential conflict. These skills would have emerged as a part of the altruistic aspects of survival in the ice age, rather than being the belated consequence of having to grapple with the consequences of warfare.

The related aspect of warfare is the contemporary male preoccupation with sport in general and team sports in particular. As the Duke of Wellington said: 'The Battle of Waterloo was won on the playing fields of Eton.' A less enamoured view comes from George Orwell, who observed that sport was war 'without the shooting.' Clearly, sport fills some of the gaps that were occupied by warfare in the past, and by extension the demands of the prehistoric hunter-gatherer lifestyle. Indeed, it could be argued that the physical demands of many sports are specifically designed to train the human body for the demands of hunting large animals on foot with spears or bows and arrows. Furthermore, the cooperative elements of team sports develop precisely the group skills that are needed for coordinated hunting activity.

In considering sport there could be a case for extending our thinking to a more general aspect of human behaviour. This is the nature of play and games in human social and intellectual development. The Dutch philosopher, Johan Huizinga, explored this concept, before the Second World War, in his book *Homo Ludens: A Study of the Play Element of Culture*. In this analysis play extends way beyond recreation and relaxation into the subtle nature of game playing in rationalising our surroundings and constructing models of how the world about us functions, in order to plan ahead.

The capacity to visualise returning herds of herbivores and to plan how to intercept and kill large numbers of prey must have been a vital part of survival. Once established, it is not hard to imagine how this skill would carry through into the other aspects of life, especially as the hunter-gatherer lifestyle is assumed to have had more leisure time. Indeed, boredom may have been one of the burdens that palaeolithic humans experienced in waiting for their prey to arrive: game-playing helped to while away the hours. What is less clear is whether this behaviour was part of the inherent consciousness of modern humans, or evolved as part of a defensive response to the trials of the ice age.

There is another aspect of game playing that may have evolved in the ice age. This is the role of humour in our creative activities. Some of the quirkier aspects of cave paintings, rock and mobiliary art may be the earliest evidence of this creative aspect of modern humans at play. For example, an exquisitely carved reindeer-antler spear thrower (*atlatl*), found in the cave of Le Mas d'Azil, Ariege, France, shows a young ibex with an emerging turd on which two birds are perched. Dating from around 16 kya, this less than reverential sense of fun obviously struck a chord at the time, as fragments of several other examples of this motif have found in this part of France. This playful approach to creativity runs through many subsequent expressions of our artistic development.

7.4 CLIMATIC DETERMINISM: THE BENEFITS OF TEMPERATE ZONES

Alongside the issues of the impact of climate change on human evolution there is the difficult question of the economic and social evolution of societies around the world since the start of the Holocene. Put simply, given that around 12 kya everybody was pretty much at the same starting point, why did certain parts of the world end up so much better off than others? In terms of the discussion in earlier chapters, are there climatic reasons why Eurasians (and peoples of Eurasian origin in the Americas and Australasia) have come to dominate the modern world in wealth and development? These questions take us into the dangerous area of climatic determinism.

The question of climatic, or more widely geographic, determinism gained widespread attention in the first half of the twentieth century. It was made famous by the writings of Ellsworth Huntingdon, a professor of geography at Yale University. The fact that this work had strong racial overtones and became linked with the discredited aspects of the subject of eugenics subsequently led to the wholesale dismissal of Huntingdon's work. Nonetheless, the underlying economic observation that the countries in the temperate parts of the world have scooped the riches of the world in recent centuries is undeniable. As a result, this whole issue has been re-examined in recent years in a manner that disposes of the racial explanations not only on moral grounds, but also on the genetic evidence of our recent shared inheritance that has been discussed at length earlier, but was unknown in the early twentieth century.

In the forefront of this work has been Jared Diamond. His book *Guns, Germs and Steel* (see Bibliography) provides a brilliant and illuminating discussion of the reasons for the long-term advantages of Eurasia in constructing economies and diffusing technologies. In particular, the spread of plant species and domesticated animals is more effective within ecological zones where climatic conditions are not dramatically different. This is because plants and animals appropriate to one ecological zone may be completely inappropriate elsewhere. Diamond argues that this spread occurs more readily in an east–west direction along the same latitude with the same day length and the same seasonal variations, rather than in a north–south direction, which almost invariably crosses ecological zones. So, despite the wide range of climatic conditions across its huge extent, Eurasia enjoyed the benefit of its vast east–west axis concentrated in temperate ecological zones.

Oceans cut off human populations in the Americas and Australasia from the vast majority of their fellows in Eurasia and Africa. They therefore could not share, through trade and diffusion, in technological advances in agriculture, communications, transport and the like. Although connected to Eurasia, Africa suffered the disadvantage of its north–south axis with its range of climates from the Mediterranean, in the north, through the superarid Saharan desert, the equatorial tropics, before returning to subtropical and Mediterranean regions in the south.

When it comes to indigenous plant and animal species, Eurasia has an even greater long-term advantage over the rest of the world. Out of about 200 000 wild plant species in the world, only a few thousand are edible, and just a few hundred have ever been domesticated. Globally there are some 50 to 60 heavily seeded wild grasses that are the most obvious candidates for plant domestication, of which nearly two-thirds grow naturally somewhere in western Eurasia, predominantly in the Mediterranean areas of the Near East. By comparison only six species grow naturally in eastern Asia, and just two in Australia and South America.

Eurasia again has all the luck when it comes to big, terrestrial, herbivorous mammals: the animals most suited to domestication. Out of 148 species of such mammals weighing more than 45 kg, only 14 have ever been domesticated. The remaining 134 have all proven impossible to domesticate for various reasons. Nine of the 14 big mammals that have been domesticated were found in the Near East, including the 'big four': the wild ancestors of goat, sheep, pig and cattle. By way of contrast, South America has a single suitable species (the llama) and North and Central America, Australia and sub-Saharan Africa have none. In the case of North America the likely candidates – house and camel – became extinct at the end of the ice age. Sub-Saharan Africa is, however, the most unfortunate because the region has 51 of the 148 heaviest mammals on Earth, but not a single one came up to scratch for domestication.

The success of Eurasia is not a matter of humans in other parts of the world failing to exploit some crops or animals that could have been successfully domesticated. Modern hunter-gatherers exhibit a detailed knowledge of every imaginable use that could be made of hundreds of plants and animals that live within their personal domain. This profound knowledge of the natural environment, gained over many generations of observation, is a vital component of the successful management of the lives of these peoples. Furthermore, when, for example, as part of the Bantu expansion (see Section 5.18) cattle spread south of the Sahara, hunter-gatherers there quickly adopted a pastoral lifestyle. Similarly, when horses were introduced into the Americas, native inhabitants immediately developed great skills in using them.

An interesting extension of the analysis of the benefits of temperate zones comes from recent work by William Masters at Purdue University and Margaret McMillan at Tufts University, which has highlighted the economic importance of climate (Masters & McMillan, 2001), as well as climate-influenced factors such as disease ecology and soil formation. In particular, the absence of winter frosts in the tropics seems to impose a significant burden inhibiting improvements in agriculture or human health. In temperate regions the regular seasonal cycle of winter frost selectively kills or enforces dormancy on pests, parasites and disease vectors. This helps raise the return on investment in human health and in crops or livestock.

Frost is particularly important for agriculture by limiting the breakdown of organic matter in soils and therefore aiding the formation of topsoil. Frozen ground also preserves moisture through the winter for release in the spring, thereby making more effective use of the available water. The influence of climate on soil formation is profound: for a given level of rainfall, organic matter breaks down about five times faster in the tropics than in temperate climates, because of greater microbial and other biotic activity. As a result, topsoil can accumulate faster in temperate than tropical regions, and temperate regions have deeper topsoil of more recent origin. In the tropics, a much larger fraction of organic matter is stored in live plants above the ground.

Human health has formed an important part of the work of Jeffrey Sachs (Gallup, Sachs & Mellinger, 1999), while Head of the Center of International Development at Harvard University, as part of his major contribution to the general debate on geography and economic development. In particular he has explored the impact of malaria on the economics of development in tropical countries, notably in Africa (Gallup & Sachs, 2000; Sachs, 2000). This is, in essence, a disease of the tropics and the subtropics, because a key part of the life cycle of the parasite (*sporogony*) depends on a high ambient temperature. Malaria also depends on adequate conditions for mosquito breeding, which is relatively clean water, usually due to rainfall forming in still pools. Malaria is endemic in the humid tropics where there is heavy rain throughout the year. But in the sub-humid tropics, where wet and dry seasons alternate, it has a distinct seasonality as mosquitoes breed during the rainy season. Additionally, the intensity of malaria transmission depends on the specific mosquito vectors that are present. Mosquitoes of the genus *Anopheles* transmit all malaria. Some anopheles species, especially those in sub-Saharan Africa, show a high preference for taking their blood meals from humans (*anthropophagy*) as opposed to animals such as cattle. This is a cruel example of where evolution has worked to the distinct disadvantage of humans, as these human-biting vectors lead to more intensive transmission of the disease.

The conclusion of Sachs and colleagues concerning the economic impact of malaria is that the location and severity of malaria are mostly determined by climate and ecology, not poverty as such. Areas with intensive malaria are almost all poor and continue to have low economic growth. The geographically favoured regions that have been able to reduce malaria have grown substantially faster since effectively eradicating the disease. The estimated impact of malaria on economic growth is large, but the mechanisms behind the impact are not clear.

In terms of economic history, the implications of tropical diseases like malaria are more difficult. Until the public health reforms of the nineteenth and twentieth centuries, which included the successful eradication of malaria in many Mediterranean countries, there was a wide range of diseases that took a dreadful toll of societies in temperate zones. So it is more difficult to establish whether the epidemic nature of many of these diseases, when combined with the relative immunity of the populace to the endemic diseases, was less economically damaging than the unremitting burden of a disease like malaria on tropical communities. The argument has to be that the continual nature of malaria with its devastating impact on infant mortality and sustained morbidity during childhood exerts a far greater economic burden on societies.

The relationship with the mosquito may go all the way back to our origins in Africa, but be heavily modified by the emergence of agriculture. Genetic mapping of the mtDNA of the malaria parasite that produces the virulent form of the disease, *Plasmodium falciparum*, from 100 sites around the world (Joy *et al.*, 2003), points to a sudden surge in the parasite population in Africa about 10 kya. This expansion was followed by a migration of the parasite to other parts of the world by about 6 kya. But the analysis of malaria parasites in South America and Asia suggest that they evolved around 100 kya. This suggests the parasite could have hitched a ride out of Africa when modern humans spread out to colonise the world between 100 and 50 kya. The spread of the disease was, however, severely restricted by climatic conditions and human behaviour during the ice age. Only with the warmth of the Holocene and the emergence of agriculture and irrigation did we create the right conditions, in the form of abundant stagnant water close to human dwellings, for the incidence of the disease to explode in its human hosts. In effect, the genetic data suggest that the ice age acted as a bottleneck for *Plasmodium falciparum* around the world, and agriculture removed the cork.

Whatever the genetic history of malaria, one thing is certain. This is that there need be no reason why this historic sense of climatic or geographic determinism should apply to the future. As places like Hong Kong and Singapore clearly show, once a state achieves both a high level of public health and sufficient wealth effectively to control its environment for its principal economic activities, it can become independent of its climatic circumstances and compete with the most successful economies in the world.

7.5 AMBIVALENCE TO ANIMALS

In Chapter 4 it was argued that the anthropomorphisation of animals was a vital component in the effective survival of hunter-gatherers in ice-age Eurasia. This empathy with animals probably also played a part in the successful domestication of animals in the Neolithic and the establishment of agriculture. The relationship has now been converted into a more ambivalent response by many of us in the modern world.

The changing relationship must have been an integral feature of the domestication of cattle, sheep and goats. Analysis of the bones of goats in Iran dating from around 10 kya provides insight into how this change occurred (Zeder & Hesse, 2000). The transition from hunting, pure and simple, to a move towards domestication took the form of a selection process of principally culling young males. This had the obvious benefit of increasing the chances of the herds continuing to breed and maintain population levels. This approach would have become more targeted by choosing to spare those animals that had the most desirable characteristics, seen most clearly in the declining size of domesticated varieties. The choice would have covered a variety of features, but would have included being responsive to domestication. As a consequence the animals that survived this selection process, and hence extinction, were those that were most amenable to a relationship, albeit lopsided, with humans.

The close relationship with dogs has already been discussed at length in Sections 3.15 and 4.10. In considering our ambivalence to animals this bond needs to be extended to the more general issue of the lengthy working arrangements we have with other animals. For example, the domestication of cats appears to date from the birth of agriculture and the storage of grain. They would have been valuable in protecting stores against rats and mice – a well-known archaeological marker of sedentism. The recent discovery that as early as 10 kya, domesticated cats were taken with settlers to Cyprus (Vigne *et al.*, 2004) is a measure of how quickly this tactic emerged as part of agriculture. The mutually beneficial nature of this arrangement cannot disguise the fact, however, that our relationship with the cat can best be defined as being 'at arm's length'.

The horse is an altogether different matter. It appears to have first been domesticated around 6 kya in Central Asia. Here the real advance was to see the horse as so much more than a source of food. This hinged on the vital discovery that it could be trained to work in close harmony with humans. Its value for transport, physical work and, most of all, for warfare has raised our relationship with this animal to the same level as that already attained with dogs. It is tempting to suggest that it was the close working relationship that hunter-gatherers in Central Asia had developed over many millennia with dogs that was the foundation for this incredibly productive development.

Nowadays most of us do not work closely with animals. Nevertheless, we have not lost our deep-seated evolutionary links with them. They are reflected in a strange mixture that often includes the anthropomorphisation not only of pets but also of cuddly toys while maintaining a studied disregard for the welfare implications of the creatures that are raised to feed us.

7.6 UPDATING OF GENDER ROLES

Having addressed the consequences of living through the ice age on the shaping of gender roles, it is now necessary to face up to the question of how this moulding process influences our views on modern society. To say the least, this opens up some dangerous ground. To start with there is the complex transition from a hunter-gatherer society to an agrarian one. One likely consequence of a sedentary life was a shortening of the time between pregnancies, which could have led to higher overall material mortality and posibly a shift in gender roles. This may have promoted the rise of male-dominated hierarchical social structures, which became the model for many societies for much of the Holocene. At the same time, the rise of organised religions has added further additional complexity to the gender roles. Then in the developed world there are the rapid transitions that have occurred since the onset of the Industrial Revolution, and with the emergence of universal suffrage.

All of these developments are beyond the scope of this book. As with other questions that have been trailed in this chapter, the only issue we can address here is whether the evolutionary imprint that is left within modern humans in surviving the ice age can illuminate our understanding of our current sociological thinking. It may be a heroic leap, but this question can then be reduced to asking whether the features identified in Section 4.7 can inform our current debate about roles of men and women in a post-modern society.

In one respect, because it can be argued that the hunter-gatherer society of the ice age was more egalitarian both in terms of individuals and genders, its evolutionary legacy equips us better for modern life. In terms of gender roles this could be helpful in both equal opportunities and in sharing social responsibilities within the family. At the same time we must not get carried away with the anthropologically seductive notion of the benefits of a Stone Age lifestyle being applied in an unalloyed form to meet current challenges. The gender-specific aspects of childbearing and the raising of children cannot be ignored. The fact that the social challenge of this vital role must have been more even greater during the ice age would seem to reinforce this politically incorrect thought. It could be argued that the decline in fertility in recent decades has reversed some of the consequences of the shift to agriculture. But it would be going too far to claim we are in some way returning to our Pleistocene inheritance. Suffice it to say, here as elsewhere, that the evolutionary consequences of surviving the ice age cast a long shadow into the present.

8 The future

> Forward, forward let us range
> Let the great world spin for ever down the ringing
> grooves of change.
>
> Alfred Lord Tennyson (1809–1892), *Locksley Hall*

Having explored the role of climate change in our past and its implications for our evolution, it seems appropriate to ask how this analysis affects thinking on current climate change. Any attempt to consider the climatic challenges of the future must draw on some prediction of what the climate will do in, say, the next century. Given the range of choices available, this does permit considerable freedom in selecting whatever model most effectively matches the line taken in this book. To deflect any accusation of bias, the best thing to do is to use the conclusions reached in the most recent report from the Intergovernmental Panel on Climate Change (the IPCC Third Assessment Report (TAR), published in 2001; see Bibliography). This contains the most complete statement of what the majority of the meteorological community considers the climate will do in the twenty-first century. Behind this façade there is, however, a wide range of uncertainties in many of the physical arguments, and hence in how global warming will affect different parts of the world. This allows plenty of choice in exploring how the impact of past climate change on human history can inform us about the challenges that lie ahead.

The IPCC predictions can be set against what happened during the last ice age and into the Holocene, and, in particular, when the climate shifted radically between these two periods. The cessation of large-scale climate change at the end of the Younger Dryas, together with the sharp drop in the level of climatic variability, provides yardsticks against which to assess the predicted changes that may occur in the twenty-first century. These past changes, it has been argued, played a fundamental part in the development of human social structures in the past 10 kyr. It follows that a radical shift in either of these aspects of our climate could cause major disruption to our modern economic and social structures. All of this has been well rehearsed in the long-running debate on global warming. But the extent to which drawing on the evidence of prehistory can illuminate the discussion has received less attention.

8.1 CLIMATE CHANGE AND VARIABILITY REVISITED

Before embarking upon predictions of the future we need to be more precise about climate change in human prehistory. We should remember the distinction between climate change and climatic variability and how these features of the climate altered between the ice age and the Holocene. As explained in Chapter 2, the nature of climatic fluctuations can be divided into more lasting forms of climate change, which may include sudden shifts, and the shorter-term variability.

The TAR addressed both of these concepts. It predicts that human activities will lead to a considerable warming in the century ahead. Although the amount of warming is subject to appreciable uncertainty, the consensus view (IPCC, 2001a, Summary for Policy Makers, p. 13.) is that the global temperature will rise by between 1.5 and 5.8 °C between 1990 and 2100. By any measure this must be regarded as a significant change in the climate, especially if the upper half of the range should prove to be correct. It would most certainly be a more substantial and rapid shift than anything that has happened since the start of the Holocene. As such it represents a huge challenge to future human activities and justifies concerted action to minimise its impact. In examining the lessons we can draw from the ice age, we are considering changes that are comparable with those that could conceivably occur in the next century.

The second stage of our analysis is more controversial. In drawing on the experience of the ice age, the implications of the climate becoming more variable in the future are an even more interesting issue. There is a widespread assumption that global warming will lead to more wild weather. Indeed, it is an equally widely accepted media-generated urban myth that the climate has become appreciably more variable in recent decades. But, as the IPCC report makes clear, the statistical evidence for an upsurge in extreme weather of late is not impressive (IPCC, 2001a, Section 2.7, pp. 155–164). Broadly speaking, it concludes that the clearest evidence of more extreme weather relates to an increase in the frequency of heavy precipitation events in the mid and high latitudes of the northern hemisphere. In addition, there has been a reduction of extreme low temperatures, with a smaller increase in the frequency of very high temperatures. It is, however, important to be clear how we define such statistics. This requires a closer examination of the concepts presented in Section 2.1.

For temperature statistics, where the climate is stationary over the period of the record, these parameters can be presented as what is often referred to as a 'bell curve'; the peak of the curve is the mean and the width of the curve is a measure of the variance (Fig. 8.1(a)). If the climate undergoes a warming without any change in the variance then the whole bell curve moves sideways. The consequence of this shift is that there are fewer cold days and more hot days, and a high probability that previous record high temperatures will be exceeded. If, however, there is an increase in variance but no change in the mean the bell curve becomes fatter and lower (Fig. 8.1(b)). The consequence is that there are more cold and more hot days and a high probability that previous records for both the coldest and hottest days will be broken. If both the mean and the variance increase then the bell curve both shifts sideways and becomes lower and fatter (Fig. 8.1(c)). The effect of this change is relatively little change in cold weather, but a big increase in hot weather and previous record high temperatures being exceeded. So, in principle, we can calculate how the incidence of extreme temperatures will change for a predicted rise in mean temperatures and a predicted change in variance.

The real challenge lies in predicting how the variance will change in the future. A detailed analysis of some 3000 meteorological

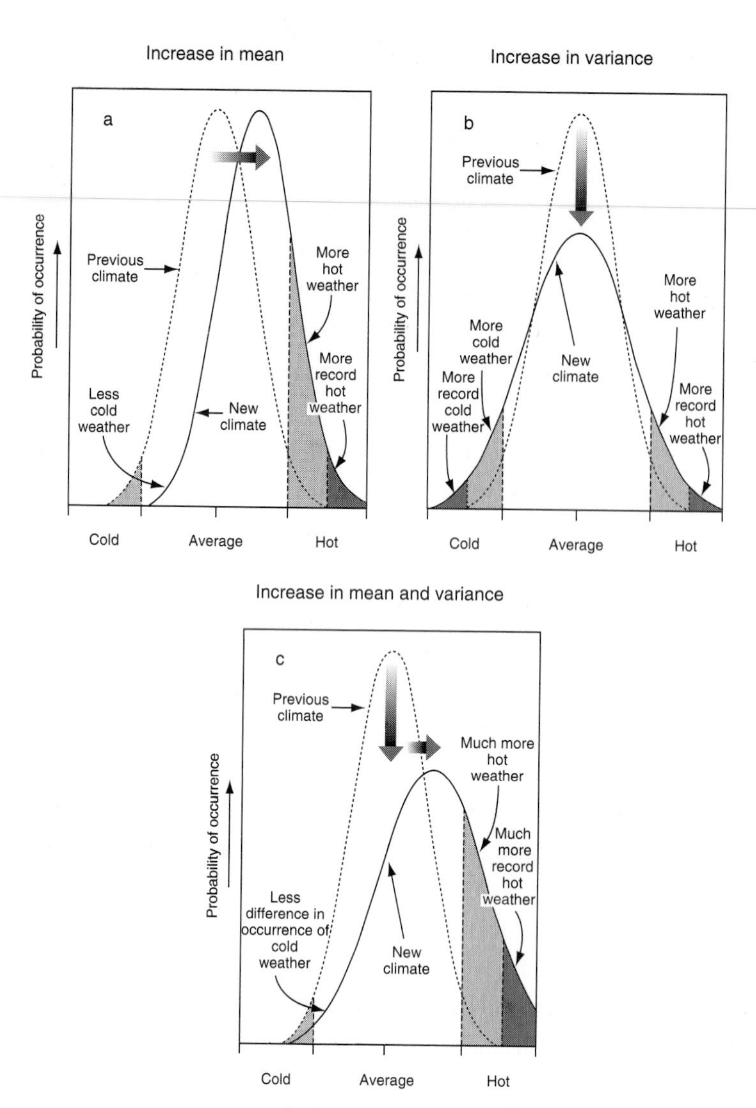

FIGURE 8.1. How climate change and shifts in climatic variability affect the incidence of extremes.

time series from sites in mid-latitudes of both the northern and southern hemispheres for the second half of the twentieth century provides some insight into how things might change in the future (Frich *et al.*, (2002). The significant temperature changes have been a decline in frost days (days with a minimum temperature less than 0 °C), a lengthening growing season, a decrease in the intra-annual temperature range, and an increase in the incidence of higher night-time temperatures, but there has been no significant increase in heatwaves. All of these observations suggest that the variability of temperature has declined. As far as rainfall is concerned, there has been a significant increase in particularly wet days and periods of wet weather, and a decline in dry periods. So, while the chances of floods have increased, the incidence of drought has not. All in all, on the basis of the past 50 years, a warmer world will feature less extreme day-to-day weather, apart from an increased risk of floods.

Such statistical analysis becomes more precarious if we extend it to cover longer-term or geographically more extensive extremes. The scale of the challenge is probably best summarised in considering predicted increases in such phenomena as warm El Niño events and tropical storms. There is an assumption that warm El Niño events will become more frequent in a warmer world. In part, this prediction is based on the fact that part of the global warming trend at the end of the twentieth century could be attributed to the major ENSO warm events in 1982/83 and 1997/98. At the same time, it is widely assumed that warmer tropical oceans will spawn more tropical storms and that, in particular, the tropical Atlantic will have more damaging hurricanes.

Examination of the climatic records steers us in the direction of a rather different set of conclusions. During the warmest period of the Holocene between 10 and 5 kya (see Section 5.19) ENSO warm events were less common. Furthermore, there is a well-established link between ENSO events and Atlantic hurricanes: the frequency of hurricanes is lowest in years when there is a strong El Niño event in the tropical Pacific. So, on the basis of the Holocene record, it is unlikely that global warming will lead to both more El Niño events and more Atlantic hurricanes. This means that it could be argued that a decline in El Niño events could lead to more Atlantic hurricanes. But this implies more cool La Niña events since hurricane frequency is greatest at times of such events. This seems unlikely as such cool events could be expected to lead to a reversal of the warming trend, given the part played by recent record-breaking El Niño events in the global warming trend of the 1980s and 1990s. Instead, the evidence of the early Holocene suggests that we will get a reduced incidence of both warm and cool events, and hence less variation from year to year in the incidence of Atlantic hurricanes.

What all this suggests is that any significant increase in the incidence of extreme weather will depend on whether the climate shifts into a regime that is markedly different from that experienced during the Holocene. This goes back to the question of whether the changes that may be wrought by human activities could lead to more sudden changes in the climate. In particular, there is the possibility that global warming could lead to a shift in the thermohaline circulation in the North Atlantic (see Section 2.11).

We will return to the question of a sudden 'flip' in the global climate in Section 8.5. For the moment, although it may be contrary to the evidence of the past 50 years, there remains the question of the statistical implications of both a significant rise in global temperature and an increase in variability of the climate. In the examples given in Figs. 2.2 and 8.1, in terms of extreme temperatures, this combination of changes equates to a warming variant of Fig. 2.2b and to Fig. 8.1c. This means that in a warmer, more variable world we would have fewer cold spells and record low temperatures, and lots more heat-waves and record highs.

Turning to rainfall, any rise in temperature is expected to pump more water vapour into the atmosphere. In itself, this change is likely to produce heavier rainfall events. If it is accompanied by a more variable climate this could lead to more floods and more droughts, depending on how seasonal weather patterns alter around the world as temperatures rise. At present, there is no clear consensus among the computer models as to how these patterns would shift. If anything, the changes might lead to wetter winters and drier summers in the mid-latitudes of the northern hemisphere, and a more intense monsoon over south Asia.

So, for the purposes of considering the future impact of both global warming and increased variability, the possibility of more erratic rainfall and higher temperatures will receive most attention.

8.2 ARE WE BECOMING MORE VULNERABLE TO CLIMATIC VARIABILITY?

There is a pervasive assumption about current climate change that we are becoming more vulnerable to extreme weather events. In terms of the subsistence crises that afflicted western Europe from the late Middle Ages through to the early nineteenth century, advances in agriculture appear to have made these a thing of the past (Burroughs, 1997, pp. 34–39). The fact that these disasters coincided with a cooler climatic period – often termed the Little Ice Age – may be an additional reason for the difficulties. But first and foremost the changes that have occurred since the late eighteenth century in western agriculture have led to a greater degree of climatic independence.

These advances have taken longer to affect the developing world. The figures for human suffering caused by the failure of the monsoon rains in India in the second half of the nineteenth century, or disastrous floods in China at the beginning of the twentieth century, are not easily comparable with more recent events (World Meteorological Organization (WMO), 2003, pp. 106–107). The mortality figures for these types of events have, however, fallen dramatically through early warning and improved disaster relief schemes. Nevertheless, droughts and floods remain the greatest climatic disasters worldwide (WMO, 2003, pp. 108–109). While droughts kill the greatest number of people, in terms of affecting the living conditions of people floods have just as great an impact.

More generally, in the developed world, the process of innovation that has been so much a part of human history has had a strong influence on our level of vulnerability to climate fluctuations. The products of this process include both major advances in agricultural productivity and industrial efficiency. In terms of our living conditions the consequent benefits range from basic domestic technologies, such as refrigeration and air conditioning, through a whole range of transportation developments from automobiles to jet aircraft, to various forms of electronic equipment which enable us to manage the complex infrastructures of modern society (WMO, 2003, p. 23). These technologies have enabled industrialised nations to construct massive networks to supply food, water, natural gas and electricity. We have also exploited climatic information to produce management and forecast systems for these infrastructures to maintain effective services. Although these systems occasionally break down in bouts of extreme weather, they have underpinned economic growth, improved public health and given us greater mobility. So, while it can be claimed with some justification that the economic costs of some forms of disaster have risen, overall much of the developed world is less vulnerable to extreme events than it was a hundred years ago.

This view could be dismissed as technological optimism. It is, however, easy to forget that many major cities in the late nineteenth century were dependent on coal for almost all of their energy supplies and nearly as reliant on horses for public transport. This dependence had serious consequences for air quality, public health and waste management. In European cities it led to serious health problems during cold spells, whereas in North America the sharpest rises in mortality occurred in summer heatwaves.

The rosy picture of our current circumstances must be treated with caution. In part, this position has come about against a background of a warming climate. If anything, this has led to slightly less variability. At the same time the benefits of technology described above may have been banked already and so there could be fewer benefits to be derived from any future developments. Nevertheless, underlying this positive view is the basic assumption that we remain an adaptable species and so will continue to respond in some way to these challenges. The same applies to the developing world where the real pressures continue to be those of rising population and the consequent pressure to live in climatically vulnerable places (e.g. floodplains and unstable hillsides). In principle, these countries have the prospect of adopting those solutions that seem to be working in the developed world, to adapt to the challenges of climate change.

8.3 CAN WE TAKE GLOBAL WARMING IN OUR STRIDE?

Even if we are adapting to current climatic variability, can we handle future shifts in the climate? The ability of countries around the world to accommodate any such changes depends both on their scale, and more crucially on how rapidly they take place. In terms of rising temperatures, the fact that humans have established thriving communities in almost every conceivable climatic zone of the globe is a measure of the adaptability of our race. This ability to accommodate such a range of extremes has, however, taken time. If global warming was to lead simply to a latitudinal shift in temperature zones, together with a relatively modest parallel adjustment in rainfall patterns, then many features of human life might be able to adjust to these changes in due course. As an indication of this ability to respond, the IPCC TAR suggests that a global rise in temperature of less than about 3 °C by 2100 would probably be manageable (IPCC, 2001b, p. 71). Beyond that life would become much more difficult. Furthermore, a significant increase in climatic variability would pose a greater threat, while sudden shifts in climate regimes might prove even more difficult to accommodate.

Predictions of how rapid the coming temperature rise will be depend to a great extent on how much of the observed global warming over the past 100 years or so can be attributed to human activities. The IPCC TAR concluded: 'In the light of new evidence and taking into account the remaining uncertainties, most of the warming over the last 50 years appears to have been due to the increase in greenhouse gas concentrations' (IPCC, 2001a, Summary for Policy Makers, p. 10). Analysis (IPCC, 2001a, pp. 134–136; Mann & Jones, 2003) of proxy records of temperature trends over the past 1800 years and instrumental records since the mid nineteenth century presents an even more definite picture (see Fig. 8.2). The shape of this time series, which has become known as the 'hockey stick', suggests that the recent warming trend dwarfs earlier natural variations over the past two millennia or so.

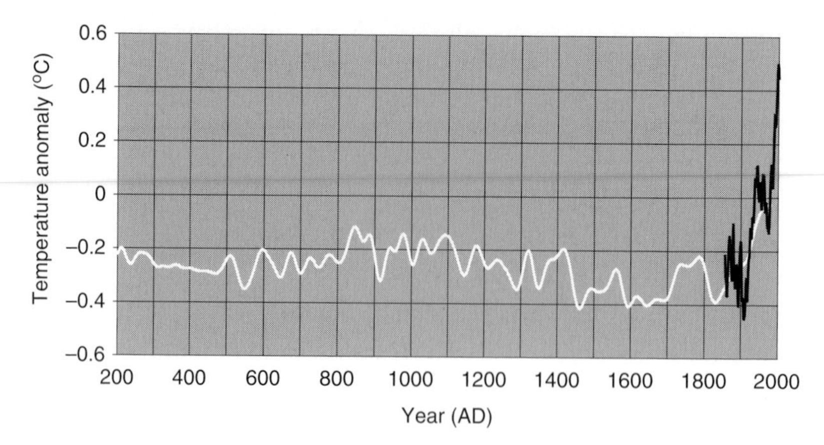

FIGURE. 8.2. A reconstruction of mean surface temperature in the northern hemisphere over the past two millennia based on high-resolution proxy temperature data (white curve), which retain millennial-scale variability, together with the instrumental temperature record for the northern hemisphere since 1856 (black curve). This combination indicates that late twentieth-century warmth is unprecedented for at least roughly the past two millennia for the northern hemisphere. (Data archived at the World Data Center for Paleoclimatology, Boulder, Colorado, USA.)

If the 'hockey stick' is an accurate assessment of climatic variations during the past two millennia, it has profound implications. Because the recent warming is so much more marked than earlier natural fluctuations, such as the Medieval Climatic Optimum and the Little Ice Age, it implies that the gentle long-term downward trend, which was the most obvious natural variation, was reversed dramatically around the end of the nineteenth century, as a result of human activities. If true, it would appear that current efforts to moderate our impact on the climate are, to say the least, a bit late.

In keeping with so many other areas covered in this book, the hockey stick is, however, the subject of intense controversy. The latest development in this particular saga is the publication of recent papers (von Storch *et al.*, 2004; Moberg *et al.*, 2005) that suggest that the methods used to generate the hockey stick underestimate past natural climatic variability. The first of these papers was greeted with headlines such as 'Hockey stick reduced to sawdust', and quotes from various quarters littered with some scientifically intemperate words about the quality of the analysis. So, while the warming since 1990 appears to exceed anything seen in the last two millennia, this looks like another debate set to run and run. In the meantime, we will have to continue to respond to observed changes within the framework of existing international agreements. If, however, the hockey stick is correct, then the real issue will be how we adapt to what appears to be an inexorable change while, at the same time, not making matters worse.

8.4 WHICH AREAS ARE MOST VULNERABLE TO INCREASED VARIABILITY?

Any significant increase in climatic variability would manifest itself in a variety of forms. Assuming that the most disruptive forms of increased fluctuations would be in terms of temperature, rainfall and windiness, it is interesting to see where might be most at risk from these changes. Working on the basis of the conclusions reached in Section 8.1, we will concentrate on the implications of higher temperatures and more erratic rainfall. Indeed, examination of this sensitivity brings out in more detail the non-linear response of social systems to extreme weather events. In short, the worst matters much more than the bad.

Starting with temperature fluctuations, there are two rather different features of extremes that matter. For high temperatures the important issue is summer heatwaves. This threat affects many cities around the world. It is, however, difficult to define precisely what constitutes dangerous levels of heat. Humans, perhaps because of their African heritage, are capable of adapting remarkably well to hot conditions. Even people who have lived all their lives in temperate regions can acclimatize to tropical conditions in a matter of weeks. It is a sudden increase in temperature, therefore, that matters most. Furthermore, it is in the big cities, where the additional effects of the urban heat island combined with industrial and vehicular pollution magnify the effects of heatwaves, that the impact of global warming may be the greatest.

The cities that are most likely to suffer adverse health effects from rising temperatures are those in the mid-latitudes and subtropics of the northern hemisphere. As a general rule, when daytime highs rise above 35 °C in the United States heat-related mortality rises. The most vulnerable cities for heat-related mortality are in the northeastern quadrant of the country, rather than the hotter cities of the southern states. In east Asia, the populous cities of China and Japan are most at risk. As a general rule, when daytime highs rise above 35 °C in these cities, the mortality rises sharply. In countries where the highest temperatures are associated with the dry season, such as India, Pakistan, the Middle East and north Africa, the threshold for human suffering tends to be around 40 °C, or even 45 °C in the hottest places. With global warming these health effects will not only increase in the hotter parts of the world but will extend to higher latitudes. The two-week record-breaking heatwave of August 2003 in France, when the excess mortality was estimated to be in excess of 11 000, is a good measure of this increasing vulnerability.

When it comes to low temperatures the fact that water freezes at 0 °C is crucial to the impact of cold weather. It means that the most sensitive regions to fluctuations in winter temperatures are those where the average temperature during winter months is close to freezing. This is the case for many populous parts of the northern hemisphere. During cold spells the temperature may drop below freezing or stay below freezing for relatively long periods. In countries within the temperate zones, there can be significant differences in severity of winters from one year to the next. Over many parts of England during the four coldest winters of the twentieth century (1917, 1940, 1947 and 1963), the average daily temperature was below freezing for more than 40 days, whereas in six of the mildest winters there were no days in which the average temperature fell below this threshold. Although the difference in the mean temperature between these two sets of extreme winters was only about 5 °C, the disruptions caused by events of the colder winters were disproportionately large.

In general, major cities in the climatic belt that experiences monthly average winter temperatures between –5 °C and +5 °C are much more likely to experience wildly differing amounts of snow and ice from year to year. In the northern hemisphere this belt covers a sizeable swathe across Europe, China, Japan and North America. Areas warmer than this, such as in countries bordering the Mediterranean and in the southern United States, rarely have below-freezing conditions for long enough (days or even weeks on end) for there to be a major problem. In colder areas, such as far northern Europe and central and northern parts of Asia and North America, snow and ice are experienced sufficiently often that adequate preparations are routinely made and ensure most fluctuations have relatively little impact. There are no large cities in the southern hemisphere that experience similar disruptive conditions during the austral winter.

In terms of rising temperatures, greater variability will mean that cold spells and near record low temperatures will still occur, but less often. This means the prospects do not look too forbidding. Places that currently experience considerable disruption during relatively rare cold spells will continue to be vulnerable. The greater sensitivity to less frequent cold snaps will partly counteract their declining occurrence. In parts of the world that experience long cold winters now, warmer conditions will probably bring significant benefits. Furthermore, because cold spells have a significant impact on mortality, the reduced loss of life, as a result of warmer winters, in some countries will more than counterbalance excess mortality caused by more summer heatwaves. In permafrost zones higher temperatures may, however, prove a substantial disadvantage, as the entire infrastructure will be vulnerable to thawing ground.

A related feature of greater variability in temperatures could be the incidence of late spring frosts. These one-off events can have a crippling effect on fruit growing, horticulture and viticulture, especially in the northern hemisphere. Even in a warmer climate, increased variability will mean that late frosts are capable of causing just as much damage. Worse still, milder winters will bring on early growth that is more vulnerable to a subsequent frost. So, even though this may occur no later than was previously normal when the climate was colder, it will do much more damage .

When it comes to fluctuations in rainfall, the possibility of greater climatic variability is a much more serious threat. As noted above, droughts and floods currently represent the greatest weather disasters around the world. An increased incidence of such extremes would make agriculture much more difficult in many parts of the world. Indeed, as noted in Chapter 2, it probably true to say that if the climate returned to the chaotic behaviour at the end of the last ice age, agriculture would only be possible in the most blessed parts of the world: modern-day refugia. The examples of the development of agriculture in the mid-Holocene in southern Mesopotamia and the Nile Valley reinforce this point. They show that even in a climatically benign period, it was only those places with the most reliable supplies of water that were capable of sustaining agriculture for centuries or millennia. Although agriculture has thrived in temperate latitudes for several millennia, the prospect of much more variable weather would pose a major challenge to maintaining modern yields.

8.5 THE THREAT OF THE FLICKERING SWITCH

The combination of greater climatic variability combined with a sudden switch in the climate is the most disturbing message from the past. The thought that a significant warming could take the global climate out of the relatively stable niche it has occupied during the Holocene is unnerving. Because the changes observed during the last ice age were more rapid and in some cases significantly greater than those predicted for the twenty-first century, they go way beyond anything experienced in recorded history.

The various theories as to what caused sudden changes such as Heinrich events and DO oscillations centre on the climate being much less stable during the ice age. This instability is usually regarded as reflecting the markedly different physical conditions that prevailed during the glacial period. Similarly, the cause of the much greater noisiness of the climate during this period may also reflect the lack of stability of the climate system. So, it can be argued that where we are heading is a different scenario and cannot be compared with the past. Nevertheless, the evidence of past erratic non-linear behaviour means that we cannot rule out entirely the possibility that future climate change could take us into back into the climatic 'long grass' (see Fig. 2.9(b)): a topsy-turvy world that poses a substantial threat to our current economic and social structures.

There is an additional reason for taking the possibility of a flip in the climate seriously. Although Heinrich events and DO oscillations appear to have been peculiar to the ice-age climate, the fact that there have been faint echoes of the DO 1500-year cycle during the Holocene (Bond *et al.*, 2001) is enough to ring warning bells. Furthermore, the way the climate of the tropics shifted swiftly around 5.5 kya with the desiccation of the Sahara and the switch in ENSO behaviour shows that even the benign Holocene has not been totally immune from such changes. So, while it may well be overdoing the catastrophe scenarios to suggest that global warming might flip the climate into changes of the scale experienced during the ice age, it is less alarmist to consider triggering changes every bit as great as have occurred during the Holocene.

The underlying question here is whether the changes that human activities may be triggering could include a significant alteration of the thermohaline circulation (THC) of the oceans, which appears to have been the cause of DO oscillations during the last ice age (see Section 2.11). There is a worrying feature lurking in most of the computer models used to predict the likely impact of human activities on the climate. What they show is a weakening of the THC and a reduction of ocean heat transport into high latitudes of the northern hemisphere (IPCC, 2001a, Summary for Policy Makers, p. 53 and 73). The reassuring aspect of these models is that, even with the reduction of heat transport, there is still warming over Europe. None of these models shows a complete shutdown of the THC by 2100.

Less reassuring is the fact that none of these models show any propensity for the climate to undergo sudden shifts. Given past evidence that such changes can occur, this stability may tell us more about the models than it does about the climate. In contrast, a number of experimental models, designed to explore abrupt regime shifts in the climate, do show a marked propensity for the climate to undergo such dramatic changes. In its most extreme form such a change could consist of effectively 'switching off' the Gulf Stream. Such a shift could, in principle, occur within a matter of years and would then have dramatic consequences for the climate of the northern hemisphere over subsequent decades (Fig. 8.3). Because such shifts are essentially chaotic, they are unpredictable. So, even if we conclude that global warming increases the possibility of a sudden shift in the climate, we have no way of telling whether it is an immediate prospect or a more distant threat. What is disconcerting is that both deep water flow measurements and satellite studies of the Gulf stream suggest that the THC in the North Atlantic has slowed down appreciably in recent decades (Dickson *et al.*, 2002; Hakkinen & Rhines, 2004).

On the other hand, the most recent ice core from Greenland (North Greenland Ice Project members, 2004) has shed new light on the end of the last interglacial between 122 and 115 kya. Earlier ice cores (GRIP and GISP2, see Section 2.3) had presented a contradictory picture of the potential suddenness of changes at this time. The latest results provide unequivocal evidence of a gradual decline in temperature with the onset of glacial conditions. This suggests that, to the extent that this transition is a useful analogue for predicting future climate change, a sudden dramatic cooling may not be on the cards.

What is even less clear is whether a sudden shift in the global climate might be accompanied by an increase in variability. While the shift in itself could well constitute sufficient challenge to economic and social stability, if it was then followed by a substantial increase in variability, the impact would be correspondingly greater. As was noted in Section 8.1, the TAR does not reach a clear conclusion on whether there is any increase in climatic variability in the past

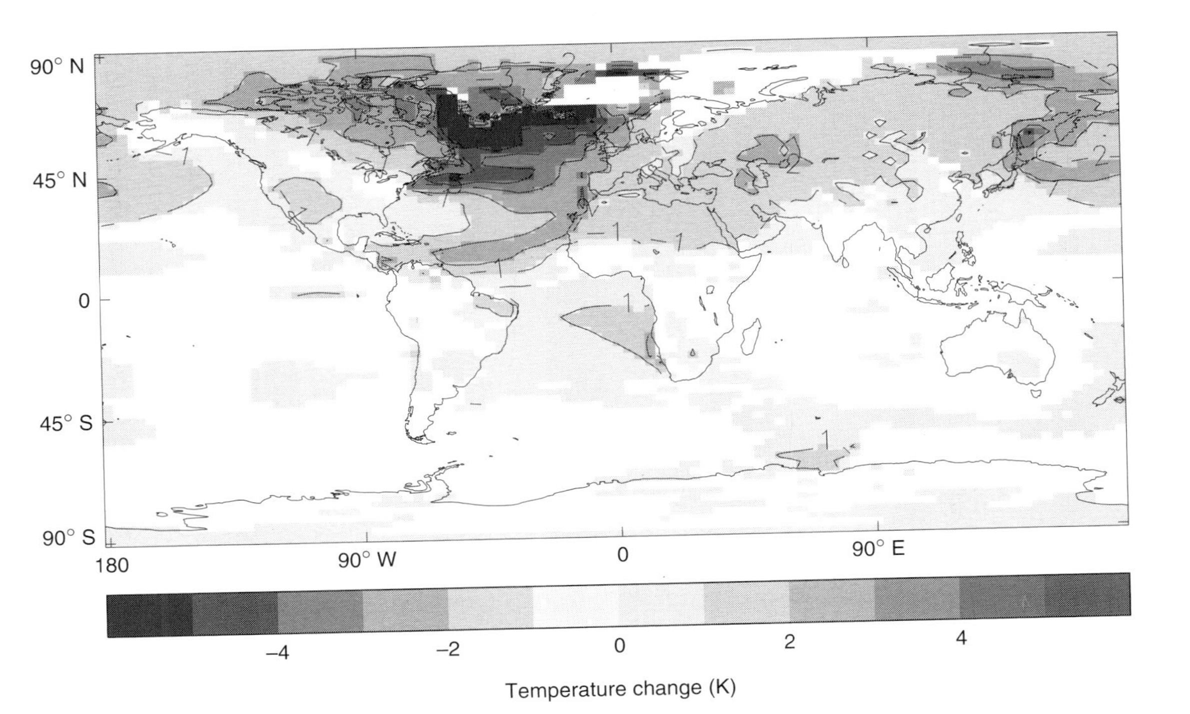

FIGURE 8.3. A computer model simulation of the changes in annual average temperatures that might occur within 50 to 100 years if the thermohaline circulation in the northern North Atlantic were to shut down. (With permission of the UKMO.)

100 years or so. So for the moment it has to be said the claims that any sudden shift in the climate will lead to an increase in extreme weather events have to be regarded as 'shroud waving'.

8.6 SUPERVOLCANOES AND OTHER NATURAL DISASTERS

While we are quaking in our beds about future climatic disasters, it is probably wise to finish off with a few words about the various natural disasters that could overtake us in the meantime. The obvious place to start is with a supervolcano. As noted in the discussion of bottlenecks (see Section 4.1), a volcano comparable in scale to Toba would have devastating effects on the world's agriculture. Massive eruptions such as Toba are estimated to occur on average about every 50 kyr or so. The most widely publicised candidate for such an eruption within the foreseeable future is Yellowstone, in the northwestern United States. This supervolcano appears to erupt roughly every 600 kyr. It last blew its top around 600 kya and in recent years has shown evidence of increased subterranean activity. The prediction, prevention and mitigation of global volcanic climatic disasters are way beyond our technological competence. So, if such explosive volcanism is the critical factor limiting the longevity of our technological society, and we can do nothing about it, then we might as well stop worrying.

Another favourite for the gloom merchants is an impact with a comet or asteroid. For one of these to have a comparable climatic impact to Toba, it would need to be of the order of 500 m across. Such an impact might be expected roughly every 100 kyr, say, a factor of two less likely than a supervolcano. Detection of an asteroid or comet of this size will probably be within the bounds of our technical skills in the next few decades. Our capacity to do anything about any such object that has our number on it is far more questionable. In these circumstances, just as with a supervolcano, it may be a case of where ignorance is bliss, 'tis folly to be wise. For those humans that survive the sort of 'bottleneck' that might occur after either a supervolcano or an asteroid impact, all they can hope to do, as their ancestors may well have done around 70 kya, is pick up the pieces and start again.

Appendix

Dating

In this book ice cores have been given pride of place in defining the timing of past climatic fluctuations. In extending the analysis to the rest of the world, and to broaden the investigation to other aspects of human prehistory, we have drawn on a wide variety of techniques for dating archaeological and geological objects and materials. These techniques have strengths and weaknesses that need to be spelt out when trying to build up a coherent picture of the past.

Experts working in a particular field will be well aware of the limitations of any particular technique they are using. But, to the outsider, the bewildering array of methods used and the apparently contradictory results they produce can lead to considerable confusion. The reasons for this complexity are not hard to find. Different methods produce different dates for events, partly because of differences in what is being measured and partly because of the inherent limitations of any given methodology.

A.1 RINGS AND LAYERS

The most reliable method of dating ancient objects, materials and events is to use either annual tree rings or undisturbed seasonal sediment layers that have been deposited at the bottom of lakes or, in special circumstances, the oceans (e.g. Cariaco Basin, Venezuela; Haug *et al.*, 2001). In the case of tree rings it has been possible to construct a series of ring widths for western Europe that extends back over 11 000 years (11 kyr; Kromer & Spurk, 1998). In the western United States studies of bristlecone pines growing high in the mountains of California have produced a series that extends nearly as far back in time (Suess & Linick, 1990). The value of these series is that not only can they used to date precisely wooden artefacts and objects found in the vicinity of bits of wood, but also they have been used to provide an independent calibration of radiocarbon dating techniques (see Section A.2 Stuiver *et al.*, 1998).

A similar set of benefits accrues from using annual lake- and ocean-sediment layers. Moreover, these records extend further back in time than tree rings. Where there is an unbroken sequence of layers whose properties contain a clear climatic signal (Haug *et al.*, 2001) it is possible to date past fluctuations, and to link these to the ice-core data that is central to the analysis in this book. In addition, these annual layers usually contain organic material that can be used for radiocarbon dating (Kitagawa & van der Plicht, 1998). So these records can also be used for calibrating radiocarbon dating farther back in time than tree rings.

In principle, the same counting of layers can be applied to ice cores. Seasonal variations in various properties of the snow (e.g. temperature-related isotope ratios and dust content) can be used to count annual layers. In practice this becomes increasingly difficult in older ice, and even in the best cores (e.g. GISP2 and GRIP) errors accumulate. Nevertheless, in these cores a variety of physical measurements has enabled scientists to detect annual variations, and estimates of age back to around 15 kya with an accuracy of a few years. Beyond this the errors increase and reach 1 to 2 kyr at 50 kya. At these great depths age profiles are usually based on models that estimate how the lower layers are compressed and extruded sideways by the mass of snow and ice on top of them. In Antarctica, where the rate of snowfall is lower, annual layers are much harder to discern and the use of flow models is an essential part of the dating process.

There is one other somewhat different aspect of layers that allows corrections to be made to dating scales. This is the existence of layers of tephra in ice cores, lake and ocean sediments that can be unequivocally attributed to specific major volcanic eruptions. The example that is cited frequently in this book is the 'supervolcano' Toba that erupted around 71 kya.

The latest analysis of speleothems, Greenland ice cores and ocean sediments (Spoetl & Mangini, 2002; Burns *et al.*, 2003; North Greenland Ice Core Project members, 2004), notably from the Cariaco basin (Hughen *et al.*, 2004) is, however, producing an increasingly coherent picture. This has led to the conclusion (Shackleton *et al.*, 2004), that, between around 29 kya and 80 kya, the dates quoted for the ice cores underestimate the age of events by between 1 and 3 kyr. This supports the case for the date of this event being around 74 kya (see Section 3.2). It does not alter the fact that wherever this layer is identified in a proxy record it marks precisely the same point in time. Furthermore, in terms of the conclusions drawn in this book, the precise dating of Toba makes no difference to the analysis of its climatic impact. What does matter is that, by identifying a number of such eruptions, it is possible to 'tune' the different records to a time-scale that at least recognises the coincidence of the markers laid down by these particular events.

A.2 RADIOACTIVE DATING

Dating rocks or preserved organic materials using their radioactive content is a simple concept. Certain widely distributed elements (e.g. carbon, potassium, thorium and uranium) occur as radioactive isotopes as well as stable isotopes. Radioactive isotopes decay. (Uranium isotopes decay to lead, potassium-40 (^{40}K) to argon-40 (^{40}Ar) and carbon-14 (^{14}C) decays to nitrogen-14 (^{14}N).) The rate of decay is characteristic for each isotope. It cannot be changed by any known force, and is defined as the *half life* – the time taken for half the number of atoms originally present to decay to daughter atoms.

In the case of rocks, provided we can assume that only the parent radioactive isotope was present when the sediments were laid down, and neither the parent nor the daughter elements have been lost or gained since the mineral was formed, it is possible to calculate its age. If the daughter element leaks away the date will be too young; if the parent leaks away the date will be too young.

Carbon-14 (^{14}C) is different. It has not been present since the Earth formed. Its half life is far too short (5770 years). Instead it is present in the atmosphere in roughly constant amount, as it is created continuously in the upper atmosphere by collisions between cosmic rays and the nuclei of nitrogen. The ^{14}C created diffuses through the atmosphere as carbon dioxide (CO_2), is dissolved in the oceans, converted by plants into organic matter, and ingested by animals. It is capable of decaying to ^{14}N as soon as it is formed, but in living organisms equilibrium is maintained with the levels in the atmosphere or the oceans. Once organisms form lasting tissue or when they die, however, the radioactive decay of ^{14}C continues without replenishment. So, provided the amount of ^{14}C in oceans and atmosphere has remained constant it is possible to use the residual radioactivity in samples of dead tissue, wood or shell to determine their age.

In practice, while the absolute measurement of the proportion of ^{14}C present in the sample has been the subject of improving experimental techniques, there are a number of sources of error in radiocarbon dating. Most significant among these is the fact that the amount of ^{14}C created in the atmosphere varies as a consequence of changes in the Sun's magnetic field. These fluctuations in ^{14}C production over the past 11 kyr can been estimated on the basis of measurements of tree rings. This is done by comparing the age estimated on the basis of constant ^{14}C production, and then comparing the difference with the known age of the tree rings and using the resultant figures to calibrate radiocarbon dates.

Beyond the range of tree rings ^{14}C-ages have been estimated using a variety of records including annually laminated lacustrine deposits (varves), where the unprecedented length of the sequence from Lake Suigetsu in Japan (Kitagawa & van der Plicht, 1998) has the potential to extend the calibration back to around 45 kya. Thus far international collaboration has produced an agreed calibration (INTCAL98; Stuiver *et al.*, 1998) that can be used to convert radiocarbon measurements into calendar years before the present back to 24 kya (Fig. A1). For the limited number of dates quoted in this book

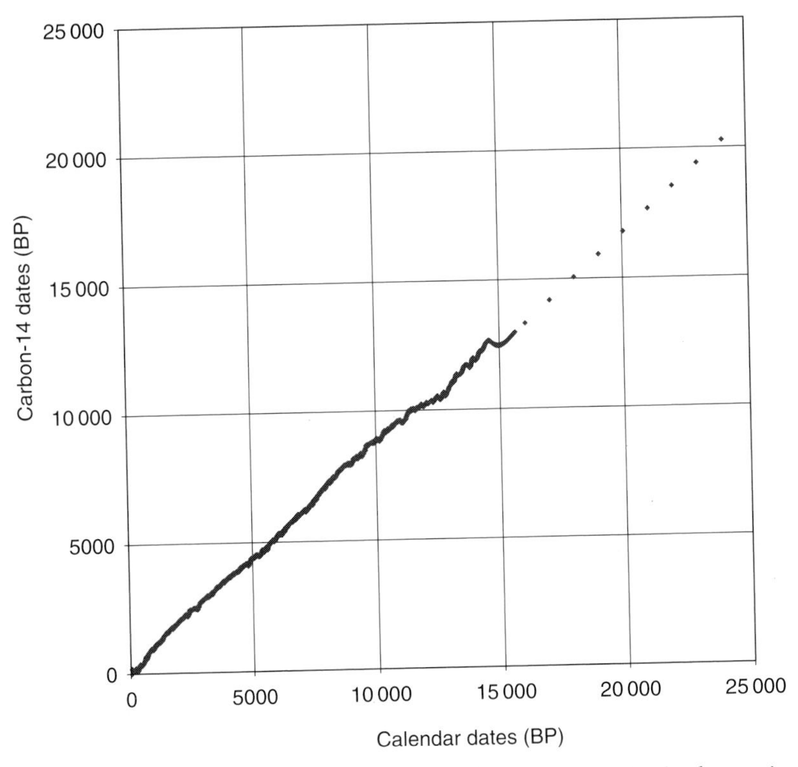

FIGURE A1. The comparison of radiocarbon dates and calendar dates using the INTCAL98 figures back to 24 kya, showing how radiocarbon dating underestimates the antiquity of organic material with increasing age.

that fall in the range 24 to 45 kya and rely on radiocarbon measurements, the calibration curve obtained from the Lake Suigetsu work is used to convert radiocarbon measurements into calendar years.

Interest in the dating process centres less on the rate at which the radiocarbaon dates diverge from the calendar dates, and more on the 'wiggles' in the calibration curve. These interfere with the precise dating of objects and materials. In addition, the most marked wiggles appear to coincide with periods of sudden climatic change, which has raised the question of a solar influence on the climate. Here, however, the real interest is in establishing a standard chronology. For ^{14}C-ages this means using the INTCAL calibration back to 24 kya. As can be seen from Fig. A1 the two start to diverge before 2 kya. By 10 kya the

^{14}C-age underestimates the real age by 1100 years, and by 15 kya the gap has risen to 2.5 kyr. It then widens more gradually to reach nearly 4 kyr at 24 kya. Beyond this the Lake Suigetsu figures bring the ^{14}C-age and the real age almost back together around 34 kya and then part a little, so that between 36 and 45 kya a standard correction of 2 kyr is added to any radiocarbon date.

The other form of isotope dating that is of particular interest in this book uses the ratio of ^{230}Th to ^{234}U in the series of decay products from uranium to lead. This method depends on finding materials that form a *closed system* in which the amount of uranium incorporated at some given time is known, and there is no external source of thorium. The calculation is complicated by ^{234}U having a half life of 245 000 years and ^{230}Th having one of 75 400 years, but this is manageable provided there was no thorium present when the sample was formed. In these circumstances, the thorium will eventually reach an equilibrium level with the uranium. The rate at which the concentration of the daughter product (^{230}Th) approaches equilibrium with the parent (^{234}U) is defined by the half life of the thorium (75 400 years).

Corals and speleothems both meet the criteria required to make them suitable for absolute dating of their formation. Living coral absorbs about 3 parts per million of uranium from seawater, and this proportion has not changed appreciably over the timescale for which corals are used in climatic studies. There is negligible incorporation of thorium into the coral. So the ratio of $^{230}Th/^{234}U$ is a measure of when the coral was formed. A similar set of physical principles applies to the measurement of the ratio of $^{230}Th/^{234}U$ in speleothems. The particular value of this technique is that it provides an absolute measure of when the coral or speleothem was formed.

A.3 BURNT FLINTS AND BROKEN TEETH

Several chronometric methods rely on the fact that an effect of naturally occurring high-energy radiation (e.g. cosmic rays) is to dislodge electrons within crystal structures. This damage accumulates over time, but can be wiped out by heating. So, the number of electrons in a substance containing crystals is a measure of the time since it was formed or last heated. It is of particular value in measuring the age of certain human remains, notably teeth, and also dating objects, such as ceramics, that were fired in their production, or the remains of hearths where flints (*silex*) have been burnt.

The most direct techniques rely on the fact that when substances are heated to around 350 °C, or stimulated optically, the free electrons in the crystal lattice emit light. These techniques are known as thermoluminescence (TL) and optically stimulated luminescence (OSL). Neither of these techniques is particularly sensitive but they can be used to date objects with ages up to several hundred thousand years, where no other alternatives are available. A more powerful technique is electron spin resonance spectroscopy (ESR). This measures the absorption of microwave radiation by unpaired electrons in paramagnetic substances. Apart from the fact that this technique is more sensitive, it has the great benefit that it does not destroy the stored information in the sample and hence can be repeated if necessary. This is particularly important where measurements are being made on human remains, such as teeth.

Independent dates obtained by ESR, OSI or TL can be set alongside other archaeological observations. It is, however, inevitable that the age results supplied by these methods are often highly dependent on the sampling strategies and circumstances within each archaeological site. Furthermore the resulting ages refer to different events as a consequence of the different methods used.

The fundamental problem to be confronted is where one method is extended to cover other artefacts that are found in association with the primary material that is dated. In case of TL-dating of silex the archaeological event that is measured is when the flint was burnt. So it is not surprising that previous ^{14}C-dating of organic material found nearby should come up with a different date. For instance, ^{14}C-sampled bones from animals that did not belong to the game of palaeolithic humans should not be linked to the human occupation. Their deposition may be totally unrelated to the time of human occupation.

Where luminescence methods are used to establish the age of the sedimentation, a different challange arises. In the case of archaeological material left by humans in palaeolithic sites, the sediment formed later. In certain circumstances these different events can be considered to be of the same age within the desired resolution of the archaeological issue being addressed. There is, however, bound to be some blurring of the dating results, depending on how long it took for the sediment to form, and this imprecision must always be kept in mind.

A.4 ADOPTING A COMMON DATING APPROACH

This brief discussion of dating techniques is designed to introduce a common format for the timescale used in this book. It recognises that the reconciliation of the different dates for many features in the prehistoric record will continue to exercise the experts for years to come. Nevertheless, because the central aim is to make the chronology of so many events and developments accessible rather than to establish an agreed set of dates, it is permissible to adopt a flexible approach. In this respect the greatest scope for confusion is in quoted radiocarbon dates, as radiocarbon measurements or calendar dates are used with almost equal frequency. So all dates are presented in the calendar form using the INTCAL98 figures to convert radiocarbon dates.

As for absolute dating, beyond the range of the tree-ring calibration of radiocarbon, and the ice-core data out to around 15 kya, we must make increasing use of Th/U dating. In particular, this is used to establish a consistent position on the major events (e.g. Heinrich layers). It is also particularly useful in making decisions between differing models of ice flow in Greenland that have led to differences between the timescales for the GISP2 and GRIP ice cores (Genty *et al.*, 2003). Combined with the tephra layers in both the ice cores and some ocean-sediment cores, it is possible to construct a sequence that underpins the template set out in Section 2.13. There are, however, times when it is easier to quote both the radiometric and the ice-core dates (e.g. in respect of the Toba eruption in Section 3.2), given the frequent use of the latter elsewhere in the book. This pragmatic approach does not undermine the attachment of particular importance to an event as the precise timing does not alter the conclusions drawn. All it does is to attempt to remove the confusing aspect of the research literature in which different studies attach a variety of dates to what is clearly a single event: better to have a single date, even if it may well be refined at some time in the future.

Glossary

Aerosols: particles, other than water or ice, suspended in the atmosphere. They range in radius from one hundredth to one ten-millionth of a centimetre. Aerosols are important as nuclei for the condensation of water droplets and ice crystals, and as participants in various atmospheric chemical reactions. Aerosols resulting from volcanic eruptions can lead to a cooling at the Earth's surface.

Albedo: the proportion of the radiation falling upon a non- luminous body that it diffusely reflects. In terms of the Earth this means the amount of sunlight that is reflected back into space by either the surface or the atmosphere.

Allerød (13.9–12.9 kya): the relatively warm period before the **Younger Dryas**.

Anatomically modern humans (*Homo sapiens*): the form of the human race that evolved in Africa between 200 and 150 **kya** and became the only representative of our species to survive the ice age. Throughout this book they are referred to as 'modern humans'.

Anticyclone: a region where the surface atmospheric pressure is high relative to its surroundings – often called a 'high'.

Aurignacian: the group of modern humans who entered Europe around 40 **kya** and became the most visible people in western Europe until the **LGM**.

Beetle assemblages: see **Coleoptera**.

Beringia: an area of land exposed during the last ice age between far eastern Siberia and Alaska that formed a land bridge between Asia and North America. This land was drowned by rising sea levels around 11 **kya**.

Blocking: a phenomenon, most often associated with stationary high pressure systems in the mid-latitudes of the northern hemisphere, which produces periods of abnormal weather.

Bølling (14.6–14.1 kya): the first major period of warming after the **Last Glacial Maximum (LGM)**.

Bottleneck: an abrupt decrease in the population of a species and, in this book, in particular, of modern humans as a result of either some global catastrophe or some more regional disaster.

Chromosomes: so-called because they are deeply stained by dyes, these rod-like structures can be seen assembling in the nucleus and then dividing equally when a cell divides into two new cells. Normally constant in number for any species, there are 22 pairs and two sex chromosomes in the human. See also **Y chromosome**.

Climate: the long-term statistical average of weather conditions. Global climate represents the long-term behaviour of such parameters as temperature, air pressure, precipitation, soil moisture, runoff, cloudiness, storm activity, winds and ocean currents, integrated over the full surface of the globe. Regional climates, analogously, are the long-term averages for geographically limited domains on the Earth's surface.

Clovis First: a term used to describe the fact that there appeared to be no occupation of North America by modern humans before those identified as being associated with bifacial stone artefacts that were first discovered at Clovis, New Mexico.

Coleoptera (beetle assemblages): the measurement of the incidence of different types of beetle in sedimentary layers that provides a measure of the climatic conditions at the time when the beetles were alive.

Continental drift: the lateral movement of continents as a result of sea-floor spreading (see also **plate tectonics**).

Dansgaard–Oeschger (DO) oscillations: sudden warmings that occurred during the last **ice age**, which appear to have been caused by periodic shifts in the circulation of the North Atlantic.

Demic Diffusion Model: an interpretation of the archaeological evidence that proposes farming was introduced across Europe by the migration of early farmers from the Middle East, rather than the ideas of agriculture being adopted by local European hunter-gatherers with relatively little replacement of the indigenous population.

Depression: a part of the atmosphere where the surface pressure is lower than in surrounding parts – often called a 'low'.

Dendrochronology: the dating of past events and variations in the environment and climate by studying the annual growth rates of trees.

Dendroclimatology: the science of reconstructing past climates from the information stored in tree trunks as the annual radial increments of growth. Wide rings signify favourable growing conditions, absence of disease and pests, and favourable climatic conditions. Narrow rings indicate unfavourable growing conditions or climate.

DNA (deoxyribose nucleic acid): in its double-stranded form the genetic material of most organisms that is faithfully copied every time a cell divides.

El Niño: an oceanic event associated with an extensive warming of sea surface temperature across the central and eastern equatorial Pacific Ocean lasting several months to more than a year.

El Niño Southern Oscillation (ENSO)**:** a quasi-periodic occurrence when large-scale abnormal pressure and sea-surface temperature patterns become established across the tropical Pacific every few years. See also **El Niño** and **La Niña**.

Epipalaeolithic: an alternative term for the **Mesolithic** period that is used to describe events in the Near and Middle East.

Feedback mechanism: a process of system dynamics in which a system reacts to amplify or suppress the effect of a force that is acting upon it. For example, in the climate system, warmer temperatures may melt snow and ice cover, revealing the darker land surface underneath. The darker surface absorbs more solar energy, causing further temperature increases, thus melting even more snow and ice cover and so on. This is positive feedback, in which warming reinforces itself. In negative feedback, a force ultimately reduces its own effect. For example, when the Earth's surface grows warmer, more water evaporates, forming more clouds. If the clouds that form are extensive and widely distributed, covering large areas of the surface, they will tend to reflect more solar radiation back into space than the dark ground underneath would, cooling the Earth's surface – and reducing the force of warmer temperatures.

Fennoscandian ice sheet: see **ice sheet**.

Foraminifera (foraminifers, or forams for short)**:** a group of microscopic marine organisms belonging to the Protozoa; they are planktonic (open-water) or benthic (bottom-dwelling) and are grazers and predators. Their calcareous shells provide a major part of the palaeoceanographic and palaeoclimatic record.

Genetic distance: a concept that expresses the fact that the measure of the genetic difference between groups of people is an indication of the time that has elapsed since the group separated: the greater the number of differences, the more ancient the separation.

Genetic drift: changes in the genetic composition of a population that arise from statistical rather than adaptive effects. Genetic drift occurs when there is a dramatic reduction in breeding numbers. Over a few generations, simple chance can lead to some genetic variations becoming widespread, or disappearing altogether. In this extreme form, this is known as *random genetic drift* or *founders' effect*.

General Circulation Models (GCMs): computational models or representations of the Earth's climate used to forecast changes in climate or weather.

Glacial: a period during the history of the Earth when there were larger ice sheets (continental-size) and mountain glaciers than today.

Glacial rebound: see **isostasy**.

Global warming: a catchall phrase to describe recent climate change some of which appears to be directly linked to human activities.

Gravettian: a group of modern humans who either entered Europe around 30 **kya** from Asia, or who were originally part of the **Aurignacian** group, but developed a separate culture that survived the **LGM** in the Balkans and eastern Europe.

Greenhouse effect: an atmospheric process in which the concentration of atmospheric trace gases (**greenhouse gases**) affects the amount of radiation that escapes directly into space from the lower atmosphere. Short-wave solar radiation can pass through the clear atmosphere relatively unimpeded. But long-wave terrestrial radiation, emitted by the warm surface of the Earth, is partially absorbed and then re-emitted by certain trace gases.

Greenhouse gases: the trace gases that contribute to the greenhouse effect. The main greenhouse gases are not the major constituents of the atmosphere – nitrogen and oxygen – but water vapour (the biggest contributor), carbon dioxide, methane, nitrous oxide and (in recent years) chlorofluorocarbons. Increases in concentrations of the latter four gases have been linked to human activity.

Hadley cell: the basic vertical circulation pattern in the tropics where moist warm air rises near the Equator and spreads out north and south and descends at around 20–30° N and S.

Half-life: time in which half of the atoms of a given quantity of radioactive nuclide undergo at least one disintegration.

Haplogroup: all those people who share a single ancestor who was the first person to possess a specific genetic mutation (usually indentified by **mtDNA** or **Y-chromosome** analysis). Such groups are often referred to by the alternative terms *gene lines*, *lineages* or *genetic groups or branches* (the definition and size of specific groups is subject to substantial variation within the genetic mapping community).

Haplotype: a set of closely linked genetic markers present on a **chromosome** or section of **DNA** that contains genetic information about only one parent (e.g. the **mtDNA** or the **Y chromosome**), which tend to be inherited together.

Heinrich events: periods of extreme cold during the last ice age when the circulation of the North Atlantic was altered by the influx of armadas of icebergs associated with the collapse of part of the **Laurentide ice sheet**.

Holocene: the relatively warm epoch that started around 10 kya and runs up to present time. It is marked by several short-lived particularly warm periods, the most significant of which, from 6.2–5.3 kya, is called the **Holocene climatic optimum**.

Holocene climatic optimum: the period of warmth in the early to mid-Holocene that marked the high point in northern hemisphere temperatures following the last **ice age**.

***Homo erectus*:** a form of archaic human that first appeared in Africa about two million years ago and spread to the warmer parts of the globe by around a million years ago.

***Homo sapiens*:** see **Anatomically modern humans**.

Hurricane: the name given primarily to tropical cyclones in the West Indies and Gulf of Mexico.

Ice ages: throughout the history of the Earth there have been times known **glacial epochs**, when there were larger ice sheets and mountain glaciers than today. The most recent glacial epoch, the **Pleistocene**, encompassed much of the last 2.5 million years. The ice age considered in this book is the most recent glacial period in the Pleistocene.

Ice sheet: a mass of land ice that is sufficiently deep to cover most of the underlying bedrock, so that its shape is mainly determined by its internal dynamics (the flow of the ice as it deforms internally and slides at its base). There are only two large ice sheets in the modern world, on Greenland and Antarctica. During the last ice age there were a number of other ice sheets, the most significant of which were the **Laurentide** (over northern North America) and the **Fennoscandian** (over northern Europe).

Insolation (from INcoming SOLar radiATION)**:** the solar radiation received at any particular area of the Earth's surface, which varies from region to region depending on latitude, weather and periodic changes in the Earth's orbit.

Interglacial: warmer periods between **ice ages** when the major ice sheets recede to higher latitudes.

Interstadial: a relatively warmer stage within an **ice age** during which temperatures rise and the advance of the ice sheets is temporarily halted or even reversed.

Intertropical Convergence Zone (ITCZ)**:** a narrow low-latitude zone in which air masses originating in the northern and southern hemispheres converge and generally produce cloudy, showery weather. Over the Atlantic and Pacific it is the boundary between the northeast and southeast trade winds. The mean position is somewhat north of the Equator but over the continents the range of motion is considerable.

Isostasy: the process whereby areas of the crust tend to float in conditions of near-equilibrium on the plastic mantle, and where ice sheets have melted this process leads to a slow rise in the crust as it returns to equilibrium (glacial rebound).

Isotopes: atoms of a single element (with the same number of protons) that have different masses (because they have a different number of neutrons). Isotopes are labelled with the approximate mass preceding the symbol of the element: ^{18}O, for example, denotes an oxygen isotope with an atomic mass of 18, instead of the mass of 16 which oxygen has under ordinary conditions (as ^{16}O). Some isotopes have characteristics making them useful for analysing chemical history; for example, they release electrons or other sub-atomic particles, allowing their presence to be detected, and they decay (gradually changing into another element) at a steady, measurable speed. The ratio between isotopes of oxygen (^{16}O and ^{18}O) can be analysed, providing a record of past surface temperatures. See also **radiocarbon dating** and **uranium/thorium dating**.

Jomon: a group of people who occupied Japan after the **LGM**, who are known for being the earliest people to develop the use of pottery.

kya: an abbreviation for 'thousand years ago'.

Lake Agassiz: the large meltwater lake that formed alongside the Laurentide ice sheet between around 13 and 8.2 **kya**, which from time to time drained into the North Atlantic via either the St Lawrence River or the Hudson Bay and triggered major shifts in the climate.

Last Glacial Maximum (LGM): the coldest period of the last ice age, which lasted from around 23 to 18 **kya**.

La Niña: an oceanic event associated with an extensive cooling of sea surface temperature across the central and eastern equatorial Pacific Ocean lasting several months to more than a year. See also **El Niño** and **El Niño Southern Oscillation**.

Laurentide ice sheet: see **ice sheet**.

Little Ice Age (AD 1550–AD 1850): a period marked by more frequent cold episodes in Europe, North America and Asia, during which mountain glaciers, especially in the Alps, Norway, Iceland and Alaska, expanded substantially.

Magdalenian: a culture that emerged after the **LGM** in southwest France and northern Iberia and reoccupied western Europe when the last **ice age** ended.

Mammoth steppe: a description of the conditions that may have prevailed across much of Eurasia during the last ice age where the combination of low precipitation, light snow cover and abundant sunshine in summer produced a biome that could support large herds of megafaunal herbivores (e.g. mammoths, woolly rhinoceros and reindeer).

Mean sea level (MSL)**:** the average height of the sea surface, based on hourly observation of the tide height on the open coast, or in adjacent waters that have free access to the sea. In the United States, MSL is defined as the average height of the sea surface for all stages of the tide over a nineteen-year period.

Medieval climatic optimum: a period of relative warmth in northern Europe between the eleventh and thirteenth centuries.

Megafauna: animals with a body weight in excess of 45 kg.

Mesolithic: the period between the **LGM** and the **Neolithic**.

Middle Palaeolithic: the middle period of the stone age, lasting from around 200 to 40 kya.

Mitochondrial DNA (mtDNA): the **DNA** contained in mitochondria that exist in the cytoplasm of human cells, which enable the cells to use oxygen to produce energy. This DNA is inherited only from the female line.

Monsoon: a seasonal reversal of wind which in the summer season blows onshore, bringing with it heavy rains, and in winter blows offshore – it is of greatest meteorological importance in southern Asia. The word is believed to be derived from the Arabic word '*mausin*', meaning a season.

Neanderthal: a form of archaic human that occupied Europe and the Middle East in the **Middle Palaeolithic** and eventually became extinct around 30 **kya**.

Neoglaciation: the first evidence of a significant advance in European mountain glaciers around 5.5 **kya**, following the warmth of the **Holocene climatic optimum**.

Neolithic: the last stage of the Stone Age, the start of which is marked principally by the adoption of agriculture. As such, its timing is dependent on regional changes in prehistoric culture.

Non-linearity: the lack of direct proportionality of the input and output of a physical system.

North Atlantic Oscillation (NAO)**:** an index of the circulation in the North Atlantic which is measured in terms of the difference in pressure between the Azores and Iceland. In winter this index tends to switch between a strong westerly flow with pressure low to the north and high in the south and the reverse: the former tends to produce above-normal temperatures over much of the northern hemisphere, the latter the reverse.

Older Dryas (14.1–13.9 kya): a brief but sharp cooling interval between the **Bølling** and the **Allerød**.

Oxygen Isotope Stages (OIS): the designated climatic stages in the standardised ocean-sediment records. There are 19 stages defining the principal **glacial** and **interglacial** periods since the Matuyama–Brunhes reversal of the Earth's magnetic field around 750 **kya**.

Palaeolithic: a term applied to the earlier part of the Stone Age of human prehistory, which is usually divide into three periods: the **lower Palaeolithic** running from the earliest evidence of human tool making, through the **middle Palaeolithic** that lasted from around 200 **kya** until about 40 **kya**, and then the **upper Palaeolithic** until after the end of the last **ice age**. See also **Epipalaeolithic**, **Mesolithic** and **Neolithic**.

Pelagic ooze: a deep ocean sediment formed from the hard parts of pelagic organisms and very fine suspended sediment.

Phenology: the study of the timing of the different stages of vegetation from year to year, covering leaf opening, flowering, fruiting and leaf fall, that can be used to establish evidence of climate change.

Phylogenetic network or tree: a diagrammatic representation of the **genetic distance** between different groups of people around the world. The more closely they are related the more recently they share a common ancestor and the shorter the branches between them, and the more remote their common ancestor the longer the branches.

Phytolith: silica deposited in the secondary plant wall of some plants, particularly grasses, which is well preserved in sedimentary deposits and can be used to identify the prevalence of specific plant types.

Phytoplankton: microscopic marine organisms (mostly algae and diatoms) which are responsible for most of the photosynthetic activity in the oceans.

Plate tectonics: the interpretation of the Earth's structures and processes (including oceanic trenches, mid-ocean ridges, mountain building, earthquake zones and volcanic belts) in terms of the movement of large plates of the lithosphere acting as rigid slabs floating on a viscous mantle.

Pleistocene (10–1600 kya): the geological period which together with the Holocene makes up the Quaternary. This epoch was characterised by numerous (at least 17) worldwide changes of climate, cycling between glacial (cool) and **interglacial** (warmer) periods, with periodicities of 100, 41 and 23 kyr.

Proxy data: any source of information which contains indirect evidence of past changes in the weather (e.g. tree rings, ice cores, corals, speleothems and ocean sediments).

Quaternary: the geological period covering the last 1.6 million years or so, which includes the **Pleistocene** and the **Holocene**.

Radiocarbon dating: a measurement technique that uses the fact that the radioactive **isotope** of carbon ^{14}C decays into ^{14}N, with a half-life of approximately 5700 years. When detected in samples of sediment or ice, isotopes can be analysed to compile records of past climate characteristics. From ^{14}C isotopes in core samples frozen in ice, the carbon content of the atmosphere over time can be derived, along with the atmospheric concentration of carbon-based molecules such as methane (CH_4).

Refugia: isolated regions in an otherwise glacial landscape where humans together with various forms of flora and fauna were able to survive the rigours of the **LGM**.

Saporels: dark, olive green to pitch black, organic-rich sedimentary layers found in ocean sediments, which provide evidence of oxygen starvation (anoxic conditions) in the bottom waters of the ocean at the time of deposition.

Solutrean: a group of humans that survived the **LGM** in the southwest of France and northern Cantabria.

Speleothems: encrustations formed by calcium carbonate deposited from running water in caves (e.g. stalactites or stalagmites). Where this encrustation accumulates at approximately constant rate, information concerning the isotopic ratios in precipitation at the time of formation can be obtained, which provides a measure of past climates.

Stadial: a colder period during the last ice age when the **ice sheets** advanced to lower latitudes.

Stratosphere: a region of the upper atmosphere, which extends from the tropopause to about 50 km above the Earth's surface, and where the temperature rises slowly with altitude. The properties of the stratosphere include very little vertical mixing, strong horizontal motions, and low water vapour content compared with the troposphere.

Sundaland: the low-lying area between Indo-China and Indonesia that was exposed by lower sea levels during much of the last ice age.

Thermohaline circulation (THC): the deep-water circulation of the oceans driven by density contrasts due to variations in salinity and temperature.

Thermoluminescence: the release of light by previously irradiated phosphors upon subsequent heating. The amount of light released is a measure of the time that has elapsed since the sample was last subjected to heating.

Time series: any series of observations of a physical variable that is sampled at constant time intervals.

Upper Palaeolithic: the most recent stage of the **Palaeolithic** that lasted from around 40 **kya** until the end of the last **ice age**.

Upper Palaeolithic Revolution: a term used to describe the apparent upsurge in creative activity that is evident in the archaeological record around 40–35 **kya** following the arrival of modern humans in Europe.

Uranium/thorium dating: a method of dating that measures the ratio of ^{230}Th to ^{234}U in geological samples containing uranium (e.g. corals and **speleothems**) that is capable of providing an absolute measure of the age of the sample.

Variance: the mean of the sum of squared deviations of a set of observations from the corresponding mean.

Varve: distinctly and finely stratified clay of glacial origin, deposited in lakes during the retreat stage of glaciation. Where these stratifications are of seasonal origin they can be used to study climatic change.

Volcanism: the phenomena of volcanic activity. Large volcanic eruptions spew massive amounts of ash into the atmosphere that absorb solar radiation, thus potentially generating a cooling effect on planetary temperatures. At the same time, volcanoes release carbon dioxide and sulphur dioxide, and the latter can produce long-lasting aerosols in the **stratosphere**.

Y chromosome: the male sex **chromosome**.

Younger Dryas (12.9–11.6 kya): a sudden, abrupt cold episode, which was the most significant interruption in the warming trend between the **Last Glacial Maximum** and the **Holocene**.

References

Aksu, A. E., Hiscott, R. N., Mudie, P. J. *et al.* (2002). Persistent Holocene outflow from Black Sea to Eastern Mediterranean contradicts Noah's flood hypothesis. *GSA Today*, **12**, 4–10.

Allen, J. R. M., Brandt, U., Brauer, A. *et al.* (1999). Rapid environmental changes in southern Europe during the last glacial period. *Nature* **400**, 740–743.

Alley, R. B., Meese, D. A., Shuman, C. A. *et al.* (1993). Abrupt increase in Greenland snow accumulation at the end of the Younger Dryas event. *Nature*, **362**, 527–529.

Alroy, J. (2001). A multispecies overkill simulation of the end-Pleistocene megafaunal mass extinction. *Science*, **292**, 1893–1896.

Ambrose, S. H. (1998). Late Pleistocene human population bottlenecks, volcanic winter, and differentiation of modern humans. *Journal of Human Evolution*, **34**, 623–651.

Ammerman, A. J. & Cavalli-Sforza, L. L. (1984) *The Neolithic Transition and the Genetics of Populations in Europe* (Princeton: Princeton Univ. Press).

Angel, J. L. (1984). Health as a crucial factor in the changes from hunting to developed farming in the eastern Mediterranean. In Cohen, M. N. & Armelagos, G. J., eds. *Paleopathology at the Origins of Agriculture* (proceedings of a conference held in 1982). (Orlando: Academic Press) pp. 51–73.

Arnold, N. S., van Andel, T. H. & Valen, V. (2002). Extent and dynamics of the Scandinavian ice-sheet during Oxygen Isotope Stage 3 (60 000 – 30 000 yr BP) *Quaternary Research*, **57**, 38–48.

Baillie, M. G. L. (1995). *A Slice Through Time: Dendrochronology and Precision Dating.* (London: Batsford).

Baillie, M. (1999). *Exodus to Arthur: Catastrophic Encounters with Comets.* (London: Batsford).

Balter, M. (2001). Did plaster hold Neolithic society together? *Science*, **294**, 2278–2281.

Bar-Yosef, O. (1986). The walls of Jericho: An alternative interpretation. *Current Anthropology*, **27**, 57–62.

(1998). The Natufian culture in the Levant: Threshold to the origins of agriculture. *Evolutionary Anthropology*, **6**, 159–179.

(2002). The Upper Paleolithic Revolution. *Annual Review of Anthropology* **31**, 363–393.

Barber, D. C., Dyke, A., Hillaire-Marcel, C. *et al.* (1999). Forcing of the cold event of 8,200 years ago by catastrophic drainage of Laurentide lakes. *Nature*, **400**, 344–348.

Barnosky, A. D., Bell, C. J., Emslie, S. D. *et al.* (2004a). Exceptional record of mid-Pleistocene vertebrates helps differentiate climatic from anthropogenic ecosystem perturbations. *Proceedings of the National Academy of Sciences*, **101**, 9297–9302.

Barnosky, A. D., Koch, P. L., Feranec, R. S., Wing, S. L. & Shabel, A. B. (2004b). Assessing the causes of late Pleistocene extinctions on the continents. *Science*, **306**, 70–75.

Bartov, Y., Goldstein, S. L., Stein, M. & Enzel, Y. (2003). Catastrophic arid episodes in the Eastern Mediterranean linked to North Atlantic Heinrich events. *Geology*, **31**, 439–442.

Blunier, T. & Brook, E. J. (2001). Timing of millennial-scale climate change in Antarctica and Greenland during the last glacial period. *Science*, **291**, 109–112.

Bond, G., Heinrich, H., Broecker, W. *et al.* (1992). Evidence for massive discharges of icebergs into the North Atlantic Ocean during the last glacial period. *Nature*, **360**, 245–249.

Bond, G. C. & Lotti, R. (1995). Iceberg discharges into the North Atlantic on millennial time scales during the last deglaciation. *Science*, **267**, 1005–1010.

Bond, G., Showers, W., Cheseby, M. *et al.* (1997). A pervasive millennial-scale cycle in North Atlantic Holocene and Glacial climates. *Science*, **278**, 1257–1265.

Bond, G., Kromer, B., Beer, J. *et al.* (2001). Persistent solar influence on North Atlantic climate during the Holocene. *Science*, **294**, 2130–2136.

Bonfils, C., de Noblet-Decoudre, N., Braconnot, P. & Joussaume, S. (2001). Hot desert albedo and climate change: Mid-Holocene monsoon in North Africa. *Journal of Climate*, **14**, 3724–3737.

Bonsall, C., Machlin, M. G., Payton, R. W. & Boroneant, A. (2002). Climate, floods and river gods: environmental change and the Meso-Neolithic transition in southeast Europe. *Before Farming*, **4**, 1–14.

Bosch, E., Calafell, F., Cornas, D. *et al.* (2001). High-resolution analysis of human Y-chromosome variation shows a sharp discontinuity and limited gene flow between Northwestern Africa and the Iberian Peninsula. *American Journal of Human Genetics*, **68**, 1019–1029.

Bowler, J. M., Johnston, H., Olley, J. M. *et al.* (2003). New ages for human occupation and climatic change at Lake Mungo, Australia. *Nature*, **421**, 837–840.

Broecker, W. S. (1995). Chaotic climate. *Scientific American*, **267**, No. 11, 44–50.

Brown, M. D., Hosseini, S. H., Torroni, A. *et al.* (1998). mtDNA haplogroup X: an ancient link between Europe/West Asia and North America? *American Journal of Human Genetics*, **63**, 1852–1861.

Burns, S. J., Fleitmann, D., Matter, A., Kramers, J. & Al-Subbary, A. A. (2003). Indian Ocean climate and an absolute chronology over Dansgaard/Oeschger events 9 to 13. *Science*, **301**, 1365–1367.

Burroughs, W. J. (1997). *Does the Weather Really Matter?* (Cambridge: Cambridge University Press).

(1998). Degrees of damage. *Journal of the Royal Horticultural Society*, **123**, 22–25.

Cambridge Encyclopaedia of Earth Sciences (1982), Smith, D. G., ed. (Cambridge: Cambridge University Press).

Cann, R. L., Stoneking, M. & Wilson, A. C. (1987). Mitochondrial DNA and human evolution. *Nature*, **325**, 31–36.

Cavalli-Sforza, L. L. (1997). Genes, peoples, and languages. *Proceedings of the National Academy of Sciences USA*, **94**, 7719–7724.

(2000). *Genes, People and Languages*. (London: Penguin Press).

Cavalli-Sforza, L. L., Menozzi, P. & Piazza, A. (1993). Demic expansions and human evolution. *Science*, **259**, 639–646.

Chikhi, L., Nichols, R. A., Barbujani, G. & Beaumont, M. A. (2002). Y genetic data support the Neolithic demic diffusion model. *Proceedings of the National Academy of Sciences USA*, **99**, 11008–11013.

Clark, P. U., Alley, R. B., Keigwin, C. D. *et al.* (1996). Origin of the first glacial meltwater pulse following the last glacial maximum. *Paleoceanography*, **11**, 563–577.

Clausen, H. B., Hammer, C. U., Hvidberg, C. S. *et al.* (1997). A comparison of the volcanic records from the Greenland Ice Core Project and Dye 3 Greenland ice cores. *Journal of Geophysical Research*, **102**, 26707–26723.

CLIMAP Project Members (1976). The surface of the ice-age Earth. *Science* **191**, 1131–1137.

CLIMAP (1981). *Seasonal reconstruction of the Earth's surface at the last glacial maximum.* Geological Society of America, Map and Chart Series, Vol. **C36**.

Clottes, J., Chauvet, J.-M., Bounel-Deschanps, E. *et al.* (1995). Les peintures paléolithiques de la Grotte Chauvet Pont d'Arc, à Vallon Pont d'Arc (Ardèche, France).*Compte Rendu de l' Academie des Sciences Paris*, **320**, 1133–1140.

Clottes, J. (2002). Palaeolithic art in France. *Adoranter, Scandinavian Society for Prehistoric Art*, pp. 5–19.

Conard, N. J. (2003). Palaeolithic ivory sculptures from southwestern Germany and the origins of figurative art. *Nature*, **426**, 830–832.

Coope, G. R., Lemdahl, G., Lowe, J. J. & Walkling, A. (1998). Temperature gradients in northern Europe during the last glacial–Holocene transition (14–9 ^{14}C kyr BP) interpreted from coleopteran assemblages. *Journal of Quaternary Science*, **13**, 419–433.

Coukell, A. (2001). Spellbound: Could mysterious figures lurking in Australian rock art be the world's oldest shamans? *New Scientist*, 19 May 2001, 34–37.

Cremaschi, M. & di Lernia, S. (1999). Holocene climatic changes and cultural dynamics in the Libyan Sahara. *African Archaeological Review*, **16**, no 4. (See also, conference abstracts *Environmental Catastrophes and Recoveries in the Holocene*. 29 August–2 September 2002, Department of Geography & Earth Sciences, Brunel University, Uxbridge, United Kingdom.)

Cruciani, F., Santolamazza, P., Sher, P. *et al.* (2002). A back migration from Asia to sub-Saharan Africa is supported by high-resolution analysis of human Y-chromosome haplotypes. *American Journal of Human Genetics* **70**, 1197–1214.

Cullen, H. M., Kaplan, A., Arkin, P. A. & deMenocal, P. B. (2002). Impact of North Atlantic Oscillation on Middle Eastern climate and streamflow. *Climate Change*, **55**, 315–338.

Dansgaard, W. & Oeschger, H. (1989). In *The Environmental Record in Glaciers and Ice Sheets*, Oeschger, H. & Langway, C. C. eds. (Chichester: Wiley), pp. 287–318.

Davis, B. A. S., Brewer, S., Stevenson, A. C. *et al.* (2003). The temperature of Europe during the Holocene reconstructed from pollen data. *Quaternary Science Reviews*, **20**, 1701–1706.

Dayton, L. (2001). Mass extinctions pinned on ice age hunters. *Science*, **292**, 1819.

deMenocal, P. D. (2001). Cultural responses to climate change during the late Holocene. *Science*, **292**, 667–673.

deMenocal, P. D., Ortiz, J., Guilderson, T. & Sarnthein, M. (2000). Coherent high- and low-latitude climate variability during the Holocene warm period. *Science*, **288**, 2198–2202.

d'Errico, F., Henshilwood, C., Lawson, G. *et al.* (2003). Archaeological evidence of the emergence of language, symbolism, and music – An alternative multi-disciplinary perspective. *Journal of World Prehistory*, **17**, 1–70.

Deser, C., Walsh, J. E. & Timlin, M. S. (1999). Arctic sea ice variability in the context of recent atmospheric circulation trends. *Journal of Climate*, **13**, 617–633.

Diamond, J. (1987). The worst mistake in the history of the human race. *Discover*, 64–66.

Dickson, B., Yashayaev, I., Meincke, J. *et al.* (2002). Rapid freshening of the deep North Atlantic Ocean over the last four decades. *Nature*, **416**, 832–837.

di Lernia, S. (2001). Dismantling dung: delayed use of food resources among early Holocene foragers of the Libyan Sahara. *Journal of Anthropological Archaeology*, **20**, 408–441.

Dillehay, T.D. (1997). *Monte Verde: A Late Pleistocene Settlement in Chile, Vol. 2: The Archaeological Context and Interpretation.* (Washington DC: Smithsonian Institution Press).

(2000). *The Settlement of the Americas: A New Prehistory*. Basic Books: New York.

Dorale, J. A., Edwards, R. L., Ito, E. & Gonzalez, L. A. (1998). Climate and vegetation history of the Midcontinent from 75 to 25 ka: A speleothem record from Crevice Cave, Missouri, USA. *Science*, **282**, 1871–1874.

Elias, S. A. (1999). Mid-Wisconsin seasonal temperatures reconstructed from fossil beetle assemblages in eastern North America: comparisons with other proxy records from the Northern Hemisphere. *Journal of Quaternary Science*, **14**, 255–262.

Emiliani, C. (1955), Pleistocene temperatures, *Journal of Geology*, **63**, 538–578.

Enzel, Y., Ely, L. L., Mishra, S. *et al.* (1999). High-resolution Holocene environmental changes in the Thar Desert, Northwestern India. *Science*, **284**, 125–128.

Eshed, V., Gopher, A., Gage, T. B., & Herkovitz, I. (2003). Has the transition to agriculture reshaped the demographic structure of prehistoric populations? New evidence from the Levant. *American Journal Physical Anthropology* **124**, 315–329.

European Project for Ice Coring in Antarctica (EPICA) Community Members (2004). Eight glacial cycles from an Antarctic ice core. *Nature*, **429**, 623–628.

Excoffier, L. & Schneider, S. (1999). Why hunter-gatherer populations do not show signs of Pleistocene demographic expansions. *Proceedings of the National Academy of Sciences USA*, **96**, 10597–10602.

Ferguson, R. B. (2003). The birth of war. *Natural History*, July–August.

Fleitmann, D., Burns, S. J., Mudelsee, M. *et al.* (2003). Holocene forcing of the Indian Monsoon recorded in a stalagmite from Southern Oman. *Science*, **300**, 1737–1739.

Florineth, D. & Schluchter, C. (1999). Alpine evidence for atmospheric circulation patterns in Europe during the Last Glacial Maximum. *Quaternary Research*, **54**, 295–308.

Foreman, S. L., Oglesby, R. & Webb, R. S. (2001). Temporal and spatial patterns of Holocene dune activity on the Great Plains of North America: megadroughts and climate links. *Global and Planetary Change*, **29**, 1–29.

Frich, P., Alexander, L. V., Della-Marta, P. *et al.* (2002). Observed coherent changes in climatic extremes during the second half of the twentieth century. *Climate Research*, **19**, 193–212.

Gagan, M. K., Ayliffe, L. K., Scott-Gagan, H., *et al.* (2002). Coral reconstruction of abrupt tropical cooling 8,000 years ago. *Geochimica et Cosmochimica Acta*, **66**, A255 (abstract).

Gallup, J. L. & Sachs, J. E. (2000). *The Economic Burden of Malaria*. Centre for International Development, Harvard University, Working Paper No 52.

Gallup, J. L., Sachs, J. E. & Mellinger, A. (1999). *Geography and Economic Development*. Centre for International Development, Harvard University, Working Paper No 1.

Gamble, C. (1999). *The Palaeolithic Societies of Europe*. (Cambridge: Cambridge University Press).

Gasse, F. (2000). Hydrological changes in the African tropics since the Last Glacial Maximum. *Quaternary Science Reviews*, **18**, 189–211.

Genty, D., Blamart, D., Ouahdi, R. *et al.* (2003). Precise dating of Dansgaard – Oeschger climate oscillations in western Europe stalagmite data. *Nature*, **421**, 833–837.

Gray, R. D. & Atkinson, Q. D. (2003). Language-tree divergence times support the Anatolian theory of Indo-European origin. *Nature*, **426**, 435–439.

Grayson, D. K. (1998). Confirming antiquity in the Americas. *Science*, **282**, 1425–1426.

Greenberg, J. & Merritt, R. (1992). Linguistic origins of Native Americans. *Scientific American*, November 1992, 94–99.

Greenberg, J., Turner, C. & Zegura, E. (1986). The settlement of the Americas: a comparison of the linguistic, dental and genetic evidence. *Current Anthropology* **27**, 477–497.

Greenland Ice Core Project (GRIP) Members (1993). Climate instability during the last interglacial period recorded in the GRIP ice core. *Nature*, **364**, 203–207.

Grimm, E. C., Jacobson, G. L., Watts, W. A. *et al.* (1993). A 50,000-year record of climate oscillations from Florida and its temporal correlation with the Heinrich events. *Science*, **261**, 198–200.

Grootes, P. M., Stuiver, M., White, J. W. C. *et al.* (1993). Comparison of oxygen isotope records from the GISP 2 and GRIP Greenland ice cores. *Nature*, **366**, 552–554.

Guiot, J. (1997). Back at the last interglacial. *Nature*, **388**, 25–27.

Guthrie, R. D. (1990). *The Frozen Fauna of the Mammoth Steppe: The Story of Blue Babe*. (Chicago: University of Chicago Press).

Haas, J. (2001). Warfare and the evolution of culture. In *Archaeology at the Millennium*, Feinmann, G. M. & Price, T. D., eds. (New York: Kluwer/Plenum). pp. 329–350.

Hakkinen, S. & Rhines, P. (2004). Decline of subpolar North Atlantic circulation during the 1990s. *Science*, **304**, 555–559.

Hanotte, O., Bradley, D. G., Ochieng, J. W. *et al.* (2002). African pastoralism: Imprints of origins and migrations. *Science*, **296**, 336–338.

Harpending, H. & Rogers, A. (2000). Genetic perspectives on human origins and differentiation. *Annual Reviews of Genomics and Human Genetics* **1**, 361–385.

Harpending, H. C., Sherry, S. T., Rogers, A. R. & Stoneking, M. (1993). The genetic structure of ancient human populations. *Current Anthropology*, **34**, 483–496.

Harpending, H. C., Batzer, M. A., Gurven, M. *et al.* (1998). Genetic traces of ancient demography. *Proceedings of the National Academy of Sciences USA*, **95**, 1961–1967.

Haas, J. (2002). Warfare and the evolution of culture. In *Archaeology at the Millennium*, Feinmann, G. & Price, D., eds. (New York, Plenum).

Haug, G. H., Hughen, K. A., Peterson, L. C. *et al.* (2001). Southward migration of the Intertropical Convergence Zone through the Holocene. *Science*, **293**, 1304–1308.

Haug, G. H., Genther, D., Peterson, L. C. *et al.* (2003). Climate and the collapse of Maya civilisation. *Science*, **299**, 1731–1735.

Heinrich, H. (1988). Origin and consequences of cyclic ice rafting in the northeast Atlantic Ocean during the past 130,000 years. *Quaternary Research*, **29**, 142–152.

Henshilwood, C. S. *et al.* (2002). Emergence of modern human behavior: Middle Stone Age engravings from South Africa. *Science*, **296**, 1278–1280.

Henshilwood, C. S. *et al.* (2004). Middle Stone Age shell beads from South Africa. *Science*, **304**, 404.

Hewitt, G. (2000). The genetic legacy of the Quaternary ice ages. *Nature*, **405**, 907–913.

Hillman, G. R., Hedges, R., Moore, A. *et al.* (2001). New evidence of late glacial cereal cultivation at Abu Hureyra on the Euphrates. *The Holocene*, **11**, 383–393.

Hodell, D. A., Curtis, J. H. & Brenner, M. (1995). Possible role of climate in the collapse of Classic Maya civilization. *Nature* **375**, 391–394.

Hodell, D. A., Kanfoush, A., Shemesh, A. *et al.* (2001). Abrupt cooling of Antarctic surface waters and sea ice extension in the South Atlantic sector of the Southern Ocean at 5000 cal yr BP. *Quaternary Research*, **56**, 191–198.

Hugen, K., Lehman, S., Southon, J. *et al.* (2004). ^{14}C activity and global carbon changes over the last 50,000 years. *Science*, **303**, 202–207.

Hurrell, J. W. (1995). Decadal trends in the North Atlantic Oscillation: Regional temperatures and precipitation, *Science*, **269**, 676–678.

Imbrie, J., Hays, J. D., Martinson, D. G. *et al.* (1984). The orbital theory of Pleistocene climate: Support from a revised chronology of the marine δ^{18}O record. In *Milankovitch and Climate*, part 1, Berger, A., Reidel, D., eds. (Massachusetts: Norwell) pp. 269–305.

Imbrie, J., Boyle, E. A., Clemens, S. C. *et al.* (1992). On the structure and origin of major glaciation cycles. 1. Linear responses to Milankovitch forcing. *Paleoceanography*, **7**, 701–738.

Imbrie, J., Berger, A., Boyle, E. A. *et al.* (1993). On the structure and origin of major glaciation cycles. 2. The 100,000 year cycle. *Paleoceanography*, **8**, 699–735.

Ingman, M. & Gyllensten, U. (2003). Mitochondrial genome variation and evolutionary history of Australian and New Guinean aborigines. *Genome Research*, **16**, 1600–1606.

Intergovernmental Panel on Climate Change (IPCC) (1990). *Climate Change: The IPCC Scientific Assessment.* Houghton, J. T., Jenkins, G. J. & Ephraums, G. G., eds. (Cambridge: Cambridge University Press).

(2001a). *Climate Change 2001: The Scientific Basis*, Houghton, J. T. *et al.*, eds. (Cambridge: Cambridge University Press).

(2001b). *Impacts, Adaptation and Vulnerabilities*, MacCarthy, J. J. *et al.*, eds. (Cambridge: Cambridge University Press).

John, P. R., Makova, K., Li, W. H. *et al.* (2003). DNA polymorphism and selection at the melanocortin-1 receptor gene in normally pigmented Southern African individuals. *Annals of the New York Academy of Sciences*, **994**, 299–306.

Joy, D. A., Feng, X., Mu, J. *et al.* (2003). Early origin and recent expansion of *Plasmodium faliciparum*. *Science*, **300**, 318–321.

Kappelman, J. (1997). They could be giants. *Nature*, **387**, 126–127.

Karlen, A. (1995). *Plague's Progress*. (London: Victor Gollancz).

Karner, D. B., Levine, J., Medeiros, B. P. *et al.* (2002). Constructing a stacked benthic $\delta^{18}O$ record. *Paleoceanography*, **17**(3), doi:10.1029/2001PA000667. (Data for this record can be found on http://jlevine.(lbl.gov/BenStack construct.html.)

Kavoutas, A., Lynch-Stieglitz, J., Marchitto, T. M. & Sachs, J. P. (2002). El Niño-like patterns in ice age tropical Pacific sea surface temperature. *Science*, **297**, 226–230.

Ke, Y. *et al.* (2001). African origin of modern humans in East Asia: a tale of 12,000 Y chromosomes. *Science*, **292**, 1151–1153.

Kirkby, P. T. (1995). Causes of short stature in coal-mining children, 1823–1850. *Economic History Review*, **xlviii**, 687–699.

Kitagawa, H. & van der Plicht, J. (1998). Atmospheric radiocarbon calibration to 45,000 yr BP: Late Glacial fluctuations and cosmogenic isotope production. *Science*, **279**, 1187–1190.

Kittler, R., Kayser, M. & Stoneking, M. (2003). Molecular evolution of *Pediculus humanus* and the origin of clothing. *Current Biology*, **13**, 1414–1417.

Klein, R. G. (2003). Whither Neanderthals? *Science*, **299**, 1525–1527.

Kremer, M. (1993). Population growth and technological change: One million BC to 1990. *Quarterly Journal of Economics*, **108**, 681–716.

Krings, M., Stone, A. , Schmitz, R. W. *et al.* (1997). Neanderthal DNA sequences and the origin of modern humans. *Cell*, **90**, 19–28.

Krings, M., Salem, A. H., Bauer, K. *et al.* (1999). mtDNA analysis of Nile River Valley populations: a genetic corridor or barrier to migration? *American Journal of Human Genetics*, **64**, 1166–1176.

Krings, M., Capelli, C., Tschentscher, F. *et al.* (2000). Neanderthal DNA sequences and the origin of modern humans. *Nature Genetics*, **26**, 144–146.

Krom, M. D., Stanley, J. D., Cliff, R. A. & Woodward, J. C. (2002). Nile River sediment fluctuations over the past 7000 yr and their key role in sapropel development. *Geology*, **30**, 71–74.

Kromer, B. & Spurk, M. (1998). Revision and tentative extension of the tree-ring based 14C calibration 9200–11, 955 cal BP. *Radiocarbon*, **40**, 1117–1125.

Kuhn, S. L., Stiner, M. C., Reese, D. S. & Gulec, E. (2001). Ornaments of the earliest Upper Paleolithic: New insights from the Levant. *Proceedings of the National Academy of Sciences USA*, **98**, 7641–7646.

Kutzbach, J. E. & Liu, Z. (1997). Response of the African monsoon to orbital forcing and ocean feedbacks in the Middle Holocene. *Science*, **278**, 440–3.

Lambeck, K. & Chappell, J. (2001). Sea level change through the last glacial cycle. *Science*, **292**, 679–686.

Lambeck, K., Esat, T. M. & Potter, E.-K. (2003). Links between climate and sea levels for the past three million years. *Nature*, **419**, 199–206.

Lell, J. T., Sukernik, R. I., Starikovskaya, Y. B., *et al.* (2002). The dual origin and Siberian affinities of Native American Y chromosomes. *American Journal of Human Genetics*, **70**, 192–206.

Leonard, J. A., Wayne, R. K. & Cooper, C. (2000). Population genetics of ice age brown bears. *Proceedings of the National Academy of Sciences USA*. **97**, 1651–1654.

Leonard, J. A. *et al.* (2002). Ancient DNA evidence for Old World origin of New World dogs. *Science*, **298**, 1613–1615.

Leuschner, D. C. & Sirocko, F. (2000). The low-latitude monsoon climate during Dansgaard–Oeschger cycles and Heinrich Events. *Quaternary Science Reviews*, **19**, 243–254.

Leverington, D. W., Mann, J. D. & Teller, J. M. (2002). Changes in the bathymetry and volume of Lake Agassiz between 9200 and 7700 ^{14}C BP. *Quaternary Research*, **57**, 244–252.

Lewis-Williams, D. (2002). *The Mind in the Cave*. (London: Thames & Hudson).

Lovvorn, M. B., Frison, G. C. & Tieszen, L. L. (2001). Paleoclimate and Amerindians: Evidence from stable isotopes and atmospheric circulation *Proceedings of the National Academy of Sciences USA*, **98**, 2485–2490.

Lu, H., Liu, Z. X., Wu, N. Q. *et al.* (2002). Rice domestication and climatic change: phytolith evidence from East China. *Boreas*, **31**, 378–385.

Maca-Meyer, N., Gonzalez, A. M., Larruga, C. *et al.* (2001). Major genomic mitochondrial lineages delineate early human expansions. *BMC Genetics*, **2**, 13.

MacPhee, R. D. E. & Marx, P. A. (1999). Mammoths and microbes: Hyperdisease attacked the New World. *Discovering Archaeology*, September/October 1999, 54–59.

Mandryk, C. A. S., Josenhans, H., Fedje, D. W. *et al.* (2001). Late Quaternary palaeoenvironments in northwestern North America: Implications for inland versus coastal migrations routes. *Quaternary Science Reviews*, **20**, 301–314.

Magny, M., Begeot, C., Guiot, J. & Peyron, O. (2003). Contrasting patterns of hydrological changes in Europe in response to Holocene climate cooling phases. *Quaternary Science Reviews*, **22**, 1589–1596.

Manley, G. (1974). Central England Temperatures: monthly means 1659 to 1973. *Quarterly Journal Royal Meteorological Society*, **100**, 389–405.

Mann, M. E. & Jones, P. D. (2003). Global surface temperatures over the past two millennia. *Geophysical Research Letters*, **30**, 1820, doi:10.1029/2003GL017814.

Manning, S. W., Kromer, B., Kuniholm, P. I. & Newton, M. W. (2001). Anatolian tree-rings and a new chronology for the east Mediterranean Bronze-Iron Ages. *Science*, **294**, 2532–2535.

(2003). Confirmation of near-absolute dating of east Bronze-Iron dendrochronology. *Antiquity*, **77**, No. 295.

Marshall, J., Kushnir, Y., Battisti, D. *et al.* (2001). North Atlantic climate variability: Phenomena, impacts and mechanisms – a review. *International Journal of Climatology*, **21**, 1863–1898.

Masters, W. & McMillan, M. (2001). Climate and scale in economic growth. *Journal of Economic Growth*, **6**, 167–186.

Mayewski, P. A., Meeker, L. D., Twickler, M. S. *et al.* (1997). Major features and forcing of high-latitude northern hemisphere atmospheric circulation using a 110,000-year-long glaciochemical series. *Journal of Geophysical Research*, **102**, 26,345–26,366.

McBrearty, S. & Brooks, A. S. (2000). The revolution that wasn't: a new interpretation of the origins of modern human behaviour. *Journal of Human Evolution*, **39**, 453–563.

Mellars, P. (1994). The Upper Palaeolithic Revolution. In *The Oxford Illustrated History of Prehistoric Europe.* Cunliffe, B., ed. (Oxford: Oxford University Press) pp. 42–78.

(2004). Neanderthals and the modern human colonisation of Europe. *Nature,* **432**, 461–465.

Meltzer, D. J. (2004). Peopling of North America, *Development in Quaternary Science,* **1**, 539–564.

Merriwether, D. A., Rothhammer, F. & Ferrell, R. E. (1995). Distribution of the four-founding lineage haplotypes in Native Americans suggests a single wave of migration for the New World. *American Journal of Physical Anthropology,* **98**, 411–430.

Miller, G. H. *et al.* (1999). Pleistocene extinction of *Genyornis newtoni*: human impact on Australian megafauna. *Science,* **282**, 205–8.

Mithen, S. J. (1994). The Mesolithic age. In *The Oxford Illustrated History of Prehistoric Europe.* Cunliffe, B., ed., (Oxford: Oxford University Press). pp. 79–135.

(2003). *After the Ice: A Global History 20,000–5000 BC.* (London: Weidenfeld & Nicholas).

Mix, A. C., Bard, E. & Schneider, R. (2001). Environmental processes of the ice age: land, oceans, glaciers (EPILOG). *Quaternary Science Reviews,* **20**, 627–657.

Moberg, A., Sonechkin, D. M., Holmgren, K., Datsenko, D. M., & Karlén, W. (2005). Highly variable Northern Hemisphere temperatures reconstructed from low- and high-resolution proxy data. *Nature,* **433**, 613–617.

Nadel, D. & Werker, E. (1999). The oldest ever brush hut plant remains from Ohalo, Jordan Valley, Israel (19,000 BP). *Antiquity,* **73**, 755–763.

North Greenland Ice Core Project members. (2004). High-resolution record of Northern Hemisphere climate extending into the last interglacial period. *Nature,* **431**, 147–155.

O'Connell, J. F. (1999). Genetics, archaeology, and Holocene hunter-gatherers. *Proceedings of the National Academy of Sciences USA,* **96**, 10,562–10,563.

Olsson, O. (2001). The rise of Neolithic agriculture. *Working Papers in Economics,* No. 57, Göteborg University.

Oppenheimer, S. (2003). *Out of Eden: The Peopling of the World.* (London: Constable).

Orlova, L. A., Kuzmin, Y. V., Stuart, A. J. & Tikhonov, A. N. (2001). Chronology and environment of wooly mammoth (*Mammuthus premigenius Blumenbach*) extinction in northern Asia. *The World of Elephants – International Congress, Rome,* pp. 718–721.

Pääbo, S. (2003). The mosaic that is our genome. *Nature,* **421**, 409–412.

Pagel, M. & Mace, R. (2004). The cultural wealth of nations. *Nature,* **428**, 275–278.

Pavlov, P., Svendson, J. I. & Indrelid, S. (2001). Human presence in the European Arctic nearly 40,000 years ago. *Nature*, **413**, 64–67.

Peterson, L. C., Haug, G. H., Hughen, K. A. & Rohl, U. (2000). Rapid changes in the hydrologic cycle of the tropical Atlantic during the last glacial. *Science*, **290**, 1947–1951.

Petit, J. R., Jouzel, J., Raynaud, D. *et al.* (1999). Climate and atmospheric history of the past 420,000 years from the Vostok ice core, Antarctica. *Nature*, **399**, 429–436.

Piperno, D. R., Weiss, E., Holst, I. & Nadel, D. (2004). Processing of wild cereal grains in the Upper Palaeolithic revealed by starch grain analysis. *Nature*, **430**, 670–673.

Pitulko, V. V., Nikolsky, P. A., Girya, E. Yu. *et al.* (2004). The Yana River site: Humans in the Arctic before the last Glacial Maximum. *Science*, **303**, 52–56.

Prentice, I. C., Jolly, D. & BIOME 6000 participants (2000). Mid-Holocene and glacial-maximum geography of the northern continents and Africa. *Journal of Biogeography*, **27**, 507–520.

Pulak, C. (1998). The Uluburun shipwreck: an overview. *International Journal of Nautical Archaeology*, **27**, 188–224.

Rahmstorf, S. (2003). Timing of abrupt climate change: a precise clock. *Geophys. Res. Lett.* **30**, No. 10, 1510, doi 10.1029/2003GLO17115.

Rahmstorf, S. & Alley, R. (2002). Stochastic resonance in glacial climate. *Eos*, **83**, 129–135.

Rampino, M. R. & Self, S. (1992). Volcanic winter and accelerated glaciation following the Toba super-eruption. *Nature*, **359**, 50–52.

Rana, B. K. *et al.* (1999). High polymorphism at the human melanocortin 1 receptor locus. *Genetics*, **151**, 1547–1557.

Reed, D. L., Smith, V. S., Hammond, S. L., Rogers, A. R. & Clayton, D. H. (2004). Genetic analysis of lice supports direct contact between modern and archaic humans. *PLoS Biology* **2** (11): e340.

Richards, M., Macauley, V., Hickey, E. *et al.* (2000). Tracing European founder lineages in the Near Eastern mtDNA pool. *American Journal of Human Genetics*, **67**, 1251–1276.

Richards, M. P., Pettitt, P. B., Stiner, M. C. & Trinkaus, E. (2001). Stable isotope evidence for increasing dietary breadth in the European mid-Upper Paleolithic. *Proceedings of the National Academy of Sciences USA*, **98**, 6528–6532.

Richards, M. P., Schulting, R. J. & Hedges, R. E. M. (2003). Sharp shift in diet at the onset of the Neolithic. *Nature*, **425**, 366.

Richerson, P. J., Boyd, R. & Bettinger, R. L. (2001). Was agriculture impossible during the Pleistocene but mandatory during the Holocene? A climate change hypothesis. *American Antiquity*, **66**, 1–50.

Roberts, R. G., Flannery, T., Aylifle, L. *et al.* (2001). New ages for the last Australian megafauna: continent-wide extinction about 46,000 years ago. *Science*, **292**, 1888–1892.

Rodbell, D. T., Seltzer, G. O., Anderson, D. M. *et al.* (1999). An ~15,000-year record of El Niño-driven alluviation in southwestern Ecuador. *Science*, **283**, 516–520.

Rogers, A. R. (2001). Order emerging from chaos in human evolutionary genetics. *Proceedings of the National Academy of Sciences USA*, **98**, 779–780.

Rohling, E. J., Fenton, M., Jorissen, F. J. *et al.* (1998). Magnitudes of sea-level low stands of the last 500,000 years. *Nature*, **394**, 162–165.

Rohling, E. J., Mayewski, P. A., Abu-Zied, R. H., Casford, J. S. L. & Hayes, A. (2002). Holocene atmosphere–ocean interactions: records from Greenland and the Aegean Sea. *Climate Dynamics*, **18**, 587–593.

Rudgley, R. (1998) *Lost Civilisations of the Stone Age* (London: Century).

Ruff, C. B., Trinkhaus, E. & Holliday, T. W. (1997). Body mass and encephalization in Pleistocene *Homo. Nature*, **387**, 173–6.

Ryan, W. B. F. & Pitman, W. (1998). *Noah's Flood. The Scientific Discoveries About the Event that Changed History* (New York: Simon and Schuster).

Ryan, W. B. F., Pitman, W. C., Major, C. O. *et al.* (1997). The abrupt drowning of the Black Sea shelf. *Marine Geology*, **138**, 119–126.

Sablin, M. V. & Khlopachev, G. A. (2002). The earliest ice age dogs: Evidence from Eliseevichi 1. *Current Anthropology*, **43**, 795–8.

Sachs, J. E. (2000). Tropical underdevelopment. *Centre for International Development, Harvard University, Working Paper No. 57.*

Sahlins, M. (1972). *Stone Age Economics* (New York: Walter De Gruyter Inc.).

Sandweiss, D. H., Maasch, K. A. & Anderson, D. G. (2001). Variations in Holocene El Nino frequencies: climate records and cultural consequences in Ancient Peru. *Geology*, **7**, 603–606.

Sarnthein, M., Pflaumann, U. & Weinelt, M. (2003). Past extent of sea ice in the northern North Atlantic inferred from foraminiferal paleotemperatures estimates. *Paleoceanography*, **18**, 1047, doi:10.1029/2002PA000771.

Savolainen, P., Zhang, Y. P., Luo, J., Lundeberg, J. & Leitner, T. (2002). Genetic evidence for an East Asian origin of domestic dogs. *Science*, **298**, 1610–1613.

Schiermeier, Q. (2004). Noah's flood. *Nature* **430**, 718–719.

Schulz, H., von Rad, U. & Erlenkeuser, H. (1998). Correlation between Arabian Sea and Greenland climate oscillations in past 110,000 years. *Nature*, **393**, 54–57.

Schulz, H., Emeis, K.-C., Erlenkeuser, H. *et al.* (2002). The Toba volcanic event and interstadial/stadial climates at the marine isotopic stage 5 to 4 transition in the Northern Indian Ocean. *Quaternary Research*, **57**, 22–31.

Schuman, B. N., Bartlein, P. J., Logar, N. *et al.* (2002). Parallel climate and vegetation responses to the early Holocene collapse of the Laurentide Ice Sheet. *Quaternary Science Reviews*, **21**, 1793–1805.

Searls, D. B. (2003). Trees of life and language. *Nature*, **426**, 391–392.

Seielstad, M. T., Minch, E. & Cavalli-Sforza, L. L. (1998). Genetic evidence for a higher female migration rate in humans. *Nature Genetics*, **20**, 278–280.

Semino, O., Passarino, G., Oefner, P. J. *et al.* (2000). The genetic legacy of paleolithic *Homo sapiens sapiens* in extant Europeans: a Y chromosome perspective. *Science*, **290**, 1155–1159.

Shelach, G. (2000). The earliest Neolithic cultures of northeast China: recent discoveries and new perspectives on the beginning of agriculture. *Journal of World Prehistory*, **14**, 363–413.

Smith, B. D. (1997). Initial domestication of *Curcurbita pepo* in the Americas 10,000 years ago. *Science*, **276**, 932–934.

Smith, P. E. L. (1976). Stone-age man on the Nile. *Scientific American*, November, p. 30.

Soffer, O., Adovasio, J. M. & Hyland, D. C. (2000). The "Venus" Figurines: textiles, basketry, gender, and status in the Upper Paleolithic. *Current Anthropology*, **41**, 511–537.

Solis, R. S., Haas, J. & Creamer, W. (2001). Dating Caral, a preceramic site in the Supe Valley on the central coast of Peru. *Science*, **292**, 723–6.

Sparks, T. H. & Carey, P. D. (1995) The responses of species to climate over two centuries: an analysis of the Marsham phenological record. *Journal of Ecology*, **83**, 321–329

Spoetl, C. & Mangini, A. (2002). Stable isotopes in speleothems as proxies to past environmental changes in the Alps. *Earth and Planetary Science Letters*, **203**, 507–518.

Stager, J. C. & Mayewski, P. A. (1997). Abrupt early to mid-Holocene climatic transition registered at the Equator and the poles. *Science*, **276**, 1834–1836.

Stanford, D. & Bradley, B. (2000). The Solutrean solution: did some ancient Americans come from Europe? *Discovering Archaeology*, **2**, 54–55.

Starikovskaya, Y. B., Sukernik, R. I., Schurr, T. G. *et al.* (1998). mtDNA Diversity in Chukchi and Siberian Eskimos: implications for the genetic history of ancient Beringia and the peopling of the New World. *American Journal of Human Genetics*, **63**, 1473–1491.

Staubwasser, M., Sirocko, F., Grootes, P. M. & Segl, M. (2003). Climate change at the 4.2 ka BP termination of the Indus valley civilization and Holocene south Asian monsoon variability. *Geophysical Research Letters*, **30**, 1425–8, doi:10.1029/2002GL016822.

Stiner, M. C. (1999). Paleolithic population growth pulses evidenced by small animal exploitation. *Science*, **283**, 190–194.

(2001). Thirty years on the 'Broad Spectrum Revolution' and paleolithic demography. *Proceedings of the National Academy of Sciences USA*, **98**, 6993–6996

Stiner, M. C. & Munro, N. D. (2002). Approaches to prehistoric diet breadth, demography and prey ranking systems in time and space. *Journal of Archaeological Method and Theory*, **9**, 181–214.

Stone, A. C. & Stoneking, M. (1998). mtDNA analysis of a prehistoric Oneota population: implications for the peopling of the New World. *American Journal of Human Genetics*, **62**, 1153–1170.

Stothers, R. B. (1984). Mystery cloud of AD 536. *Nature*, **307**, 344–345.

Street-Perrott, F. A., Holmes, J. A., Waller, M. P. *et al.* (2000). Drought and dust deposition in the West African Sahel: a 5500-year record from Kajemarum Oasis, northeastern Nigeria. *Holocene*, **10**, 293–302.

Stringer, C. (2000). Coasting out of Africa. *Nature*, **405**, 24–27.

Stringer, C. & Gamble, C. (1993). *In Search of the Neanderthals* (London: Thames & Hudson).

Stuart, A. J., Kosintsev, P. A., Higham, T. F. G. & Lister, A. M. (2004). Pleistocene to Holocene extinction dynamics in giant deer and woolly mammoth. *Nature*, **431**, 684–689.

Stuiver, M., Reimer, P. J., Bard, E. *et al.* (1998). INTCAL98 Radiocarbon age calibration, 24,000–0 cal BP. *Radiocarbon*, **40**, 1041–1083.

Suess, H. E. & Linick, T. W. (1990). The C record in bristlecone pine wood of the past 8000 years based on the dendrochronology of the late C. W. Ferguson. *Philosophical Transactions of the Royal Society of London*. **A 330**, 403–412.

Sykes, B. (2002). *The Seven Daughters of Eve* (London: Bantam Press) pp. 167–184.

Taçon, P. & Chippendale, C. (1994). Australia's ancient warriors: changing depictions of fighting in the rock art of Arnhem Land, NT. *Cambridge Archaeological Journal*, **4**, 211–248.

Taçon, P. S. C. & Pardoe, C. (2002). Dogs make us human. *Nature Australia* **27**, 52–61.

Tambets, K., Rootsi, S., Kivisild, T. *et al.* (2004). The western and eastern roots of the Saami – the story of genetic 'outliers' told by mitochondrial DNA and Y chromosomes. *American Journal of Human Genetics* **74**, 661–682.

Taylor, K. C., Lamorey, G. W., Doyle, G. A. *et al.* (1993). The 'flickering switch' of late Pleistocene climate change. *Nature*, **361**, 432–436.

Thompson, L. G., Davis, M. E., Mosley-Thompson, E. *et al.* (1998). A 25,000 year tropical climate history from Bolivian ice cores. *Science*, **282**, 1858–1864.

Thorpe, I. J. N. (2003). Anthropology, archaeology and the origin of warfare. *World Archaeology*, **35**, 145–165.

Thouveny, N., de Beaulieu, J. L., Bonifay, E. *et al.* (1994). Climate variations in Europe over the past 140 kyr deduced from rock magnetism. *Nature*, **371**, 503–506.

Torroni, A., Schurr, T. G., Cabell, M. F. *et al.* (1993). Asian affinities and continental radiation of the four founding Native American mtDNAs. *American Journal of Human Genetics*, **53**, 563–590.

Torroni, A., Bandelt, H. J., D'Urbano, L. *et al.* (1998). mtDNA analysis reveals a major Late Paleolithic population expansion from southwestern to northeastern Europe. *American Journal of Human Genetics*, **62**, 1137–1152.

Trinkhaus, E. & Shipman, P. (1993). *The Neanderthals, Changing the Image of Mankind* (London: Jonathan Cape).

Tudhope, A. W., Chilcott, C. P., Mc Culloch, M. T., *et al.* (2001). Variability in the El Niño–Southern Oscillation through a glacial–interglacial cycle. *Science*, **291**, 1511–1517.

Tzedakis, P. C., Lawson, I. T., Frogley, M. R. *et al.* (2002). Buffered tree population changes in a quaternary refugium: evolutionary implications. *Science*, **297**, 2044–2047.

Underhill, P. A., Shen, P., Lin, A. A. *et al.* (2000). Y chromosome sequence variation and the history of human populations. *Nature Genetics*. **26**, 358–361.

Usher, D. (2003). *Political Economy* (Oxford: Blackwell).

Valladas, H., Clottes, J., Geneste, J. M. *et al.* (2001). Evolution of prehistoric cave art. *Nature*, **413**, 479.

Vallois, H. (1961). The social life of early man: The evidence of the skeletons. In *Social Life of Early Man*, Washburn, S., ed. (Chicago: Aldine).

van Engelen, A. F. V. & Nellestijn, J. W. (1995). Monthly, seasonal and annual means of the air temperature in tenths of centigrades in de Bilt, Netherlands, 1706–1995. *KMNI (Dutch Meteorological Service) Report*, Climatological Services Division.

van Loon, H. & Rogers, J. C. (1978). Seesaw in winter temperatures between Greenland and Northern Europe. 1. General description. *Monthly Weather Reviews*, **106**, 296–310.

van Noordwijk, A. J. (2003). The earlier bird. *Nature*, **422**, 29.

Vartanyan, S. L., Garutt, V. E. & Sher, A. V. (1993). Holocene dwarf mammoths from Wrangel Island in the Siberian Arctic. *Nature*, **362**,337–340.

Vasil'ev, S. A., Kuzmin, Y. V., Orlova, L. A. & Dementiev, V. N. (2002). Radiocarbon-based chronology of the Paleolithic in Siberia and its relevance to the peopling of the New World. *Radiocarbon*, **44**, 503–530.

Vigne, J. D., Guilaine, J., Debue, K. *et al.* (2004). Early taming of the cat in Cyprus. *Science*, **304**, 259.

Vila, C., Savolainen, P., Maldonado, J. E. *et al.* (1997). Multiple and ancient origins of the domestic dog. *Science*, **276**, 1687–1689.

Vila, C., Maldanado, E. & Wayne, R. K. (1999). Phylogenic relationships, evolution, and genetic diversity of the domestic dog. *Journal of Heredity*, **90**, 71–77.

Verrelli, B. C. & Tishkoff, S. A. (2004). Signatures of selection and gene conversion associated with human color vision variation. *American Journal of Human Genetics*, **75**, 363–375.

von Grafenstein, U., Erlenkeuser, H., Braues, A. *et al.* (1999). A mid-European decadal isotope-climate record from 15,500 to 5000 years BP. *Science*, **284**, 1684–1687.

von Storch, H., Zorita, E., Jones, J., Dimitriev, Y., González-Rouco, F. & Tett, S. (2004). Reconstructing past climate from noisy data. *Science*, **306**, 679–683.

Walker, G. T. (1928). World Weather. *Quarterly Journal Royal Meteorological Society*, **54**, 79–87.

Wall, J. D. & Przeworski, M. (2000). When did the human population size start increasing? *Genetics*, **155**, 1865–1874.

Wallace, D. C. & Torroni, A. (1992). American Indian prehistory as written in the mitochondrial DNA: a review. *Human Biology*, **64**, 403–416.

Walter, R. C., Buffler, R. T, Bruggemann, J. H. *et al.* (2000). Early human occupation of the Red Sea coast of Eritrea during the last interglacial. *Nature*, **405**, 65–69.

Wang, Y. J., Cheng, H., Edwards, R. L. *et al.* (2001). A high-resolution absolute-dated Late Pleistocene monsoon record from Hulu Cave, China. *Science*, **294**, 2345–2348.

Weaver, A. J., Saenko, O. A, Clark, P. U., & Mitrovica, J. X. (2003). Meltwater pulse 1A from Antarctica as a trigger of the Bølling/Allerød warm interval. *Science*, **299**, 1709–1713.

Weiss, H. & Bradley, R. S. (2001). What drives societal collapse? *Science*, **291**, 609–610.

Weiss, H., Courty, M. A., Wetterstrom, W. *et al.* (1993). The genesis and collapse of third millennium North Mesopotamian civilization. *Science*, **261**, 995–1004.

Wells, R. S., Yuldasheva, N., Ruzibakiev, R. *et al.* (2001). The Eurasian heartland: a continental perspective on Y-chromosome diversity. *Proceedings of the National Academy of Sciences USA*, **98**, 10244–10249.

Wendorf, F. (1968). A Nubian Final Paleolithic graveyard near Jebel Sahaba, Sudan. In *The Prehistory of Nubia* (Dallas: Fort Burgwin Research Center and Southern Methodist University Press) pp. 791–953

Wendorf, F. & Schild, R. (1998). Nabta Playa and its role in northeastern African prehistory. *Journal of Anthropological Archaeology*, **17**, 97–123.

Wendorf, F., Schild, R., El Hadidi, N. *et al.* (1979). The use of barley in the Egyptian Late Paleolithic. *Science*, **205**, 1341–1347.

Wendorf, F., Schild, R., Close, A. E. *et al.* (1984). New radiocarbon dates on the cereals from Wadi Kubbaniya. *Science*, **225**, 645–646.

Wenke, R. J. (1999). *Patterns in Prehistory: Humankind's First Three Million Years*, 4th Edn. (Oxford: Oxford University Press).

White, T. D., Asfaw, B., De Gusta, D. *et al.* (2003). Pleistocene *Homo Sapiens* from Middle Awash, Ethiopia. *Nature*, **423**, 742–747.

Whitlock, C. & Bartlein, P. J. (1997). Vegetation and climate change in northwestern America during the last 125 kyr. *Nature*, **388**, 57–61.

Wick, L., Lemcke, G. & Sturm, M. (2003). Evidence of late glacial and Holocene climatic change and human impact in eastern Anatolia: high-resolution pollen, charcoal, isotopic and geochemical records from the laminated sediments of Lake Van, Turkey. *The Holocene*, **13**, 665–675.

Wilson, A. C. & Cann, R. L. (1992). The recent African genesis of humans. *Scientific American*, April 1992, pp. 68–73.

Wilson, J. F. *et al.* (2001). Genetic evidence for different male and female roles during cultural transitions in the British Isles. *Proceedings of the National Academy of Sciences USA*, **98**, 5078–5083.

World Meteorological Organisation (WMO) (2003). *Climate: Into the 21st Century*, Burroughs, W. J., ed. (Cambridge: Cambridge University Press).

Yu, G., Chen, X., Ni, J. *et al.* (2000) Palaeovegetation of China: a pollen data-based synthesis for the mid-Holocene and last glacial maximum. *Journal of Biogeography*, **27**, 635–664.

Yu, Z. & Eicher, U. (1998). Abrupt climate oscillations during the last deglaciation in central North America. *Science*, **282**, 2235–2238.

Zazula, G. D., Froese, D. G., Schweger, C. E. *et al.* (2003). Ice-age steppe vegetation in east Beringia. *Nature*, **423**, 603.

Zeder, M. A. & Hesse, B. (2000). The initial domestication of goats (*Capra hircus*) in the Zagros mountains 10,000 years ago. *Science*, **287**, 2254–2257.

Bibliography

Alley, R. B. *The Two-Mile Time Machine: Ice Cores, Abrupt Climate Change and Our Future*. Princeton: Princeton University Press, 2000.

A vivid description by one of the foremost researchers in the field of ice-core studies of the challenges of extracting information from the world's major ice sheets, the invaluable information this can provide about past climate change and the potential implications this has for predicting future climatic developments.

Baillie, M. G. L. *A Slice Through Time: dendrochronology and precision dating* London: Batsford, 1995.

A highly accessible presentation of the basic aspects of dendrochronology that provides an easy introduction to just how much information can be extracted from tree rings.

Baillie, Mike *Exodus to Arthur: Catastrophic Encounters with Comets* London: Batsford, 1999.

An intriguing and thorough analysis of the evidence that can be interpreted as showing that the Dark Ages may have been precipitated by climatic events after the Earth was hit by cometary fragments.

Barry, R. G. & Chorley, R. I. *Atmosphere, Weather and Climate* London: Routledge, 1998.

The seventh edition of a well-established widely read standard work which provides an up-to-date treatment of current meteorological theory and practice with a global perspective.

Burroughs, W. J. (1997). *Does the Weather Really Matter?* Cambridge: Cambridge University Press.

A book that seeks to provide a balanced and accessible analysis of the current debate on climate change and the past human activities play in current changes. It combines a historical perspective, economic and political analysis together with climatological explanations of the impact of extreme weather events on aspects of society.

Burroughs, W. J. (2001). *Climate Change: A Multidisciplinary Approach* Cambridge: Cambridge University Press.

This book provides a concise, up-to-date presentation of our current knowledge of climate change and its implications for society. This enables the reader to put the claims about weather cycles into the context of our current understanding of the causes of climate change, both in terms of current changes and those that have occurred throughout the Earth's history.

Burroughs, W. J. *Weather Cycles: Real or Imaginary?* 2nd Edn, Cambridge: Cambridge University Press, 2003.

This book explores the unresolved debate on the existence of weather cycles. It examines the competing arguments for observed effects being due to natural variability, solar activity and the Earth's orbital parameters. As such it provides the detailed background to the various aspects of apparently periodic behaviour in the climate that are needed in understanding contemporary changes in the Earth's climate.

Cavalli-Sforza, L. L. *Genes, People and Languages* London: Allen Lane: The Penguin Press, 2000.

An illuminating and personal account of the development of many of the techniques of genetic mapping by the great pioneer in the subject. It provides a fascinating set of insights by a towering figure in the discipline into how the nature of human diversity has been unravelled and a unified perspective of our history established. It also contains a valuable analysis of glottochronology.

Cunliffe, B. (ed.) *The Oxford Illustrated History of Prehistoric Europe* Oxford: Oxford University Press.

An excellent grounding in many aspects of prehistoric developments in Europe. The early chapters by Clive Gamble, Paul Mellars and Steven Mithen are particularly valuable in setting the scene for assessing the impact of climate change.

Diamond, J. *Guns, Germs, and Steel: The Fates of Human Societies* New York: Norton, 1997.

A seminal study that presents in an accessible form the various arguments for why societies developing in the temperate regions of the northern hemisphere had everything going for them.

Diaz, H. F. & Markgraf, V. (eds.) *El Niño and the Southern Oscillation* Cambridge: Cambridge University Press, 2000.

A comprehensive set of papers, which provide a useful compendium of knowledge about the part played by ENSO in the global climate and its impact on social and economic issues around the world. This analysis includes work on variations over recent centuries, but does not extend back to the changes in ENSO throughout prehistory that are mentioned in this book.

Fagan, B. *The Long Summer: How Climate Changed Civilisation* New York: Basic Books, 2004.

A different perspective on the implications of the climatic amelioration after the LGM and how climate change moulded many aspects of human history. This analysis is particularly valuable in providing a respected archaeologist's observations on how vulnerable human societies have been to shifts in the climate.

Gamble, C. *The Palaeolithic Societies of Europe* Cambridge: Cambridge University Press, 1999.

A comprehensive and accessible analysis of many features of stone-age social structures and technologies in Europe. It is of particular value in considering the background to the changes that occurred with transition from the Middle to the Upper Palaeolithic.

Grove, J. M. *The Little Ice Age* London: Methuen, 1988.

An immensely thorough review of the evidence and consequences of the Little Ice Age. It is particularly valuable in that it extends its comprehensive analysis to cover the contraction and expansion of glaciers around the world throughout the Holocene, which provides detailed information about changes in the climate since the last ice age.

Haynes, G. *The Early Settlement of North America: The Clovis Era* Cambridge: Cambridge University Press, 2002.

A particularly useful review of the complex weave of argument surrounding the peopling of the New World and the role of modern humans in mass extinctions there. Also, the author's expertise in modern ecological studies of elephants in Africa brings a more practical edge to the discussions of both the archaeological record and the nature of hunter-gatherer activities. This provides insights into the competing impacts of population growth, animal responses and climate change on the dramatic ecological shifts that occurred after the last ice age in North America.

Imbrie, J. & Imbrie, K. P. *Ice Ages: Solving the Mystery* London: Macmillan, 1979.

An accessible presentation of the research into the causes of the Ice Ages. It is particularly interesting in providing a personal insight into the work during the 1960s and 1970s that established the modern theory of the ice ages.

Intergovernmental Panel on Climate Change. *Climatic Change: The Scientific Basis* Cambridge: Cambridge University Press, 2001.

The most comprehensive set of surveys of the evidence for climate change and a detailed analysis of the consequences of anthropogenic activities, including increasing the level of 'greenhouse gases' in the atmosphere, plus forecasts of the likely impact on the global climate over the next century. In spite of the categorical statements made about contribution of anthropogenic activities to current global warming, the body of this massive and authoritative study contains a bewildering array of caveats about the uncertainties in modelling the global climate, which provide good reason for being cautious about the precise extent and nature of future warming.

Intergovernmental Panel on Climate Change. *Impacts, Adaptation and Vulnerabilities*. Cambridge: Cambridge University Press, 2001.

This is the companion volume to the technical report described above. It provides a comprehensive analysis of the evidence that recent observed changes in climate have already affected a variety of physical and biological systems. It then assesses the potential impact of future climate change on human populations, natural environments and wildlife. It considers how adaptation might lessen adverse impacts or enhance beneficial impacts.

Karlen, A. *Plague's Progress* London: Victor Gollancz, 1995.

A riveting history of the diseases that have been part of the human condition since we first started to live in villages and larger settlements. Full of wonderful images and penetrating analysis of the Faustian bargain the human race entered when it settled down to live together in large numbers and in close proximity to domesticated animals.

Lamb, H. H. *Climate – Present, Past and Future*, Vols. 1 and 2 London: Methuen, 1972, 1977.

The classic work on all aspects of climatic change, which considers the complete range of meteorology, climatology, the evidence of climatic change and possible explanations of observed changes. Of particular interest is that these works provide an unparalleled review of the wide range of work devoted to the subject of climate change that was conducted in the first half of the twentieth century. Much of this work has now been overtaken by more recent studies, but it forms an essential background to in maintaining an adequate sense of perspective in addressing the current debate about the nature and extent of current changes.

Lewis-Williams, D. *The Mind in the Cave* London: Thames & Hudson, 2002.

A comprehensive analysis of rock and cave paintings around the world by the leading authority on the subject. It contains not only a thorough description of the many works but also an idiosyncratic set of interpretations of the meaning of the images.

Mithen, S. *After the Ice: A Global History 20,000–5000 BC* London: Weidenfeld & Nicholas, 2003.

An eclectic and personalised examination of the archaeological evidence of human activities around the world during the last stages of the ice age and the first half of the Holocene. By taking a detailed tour of many famous archaeological sites and using a 'fly-on-the-wall' technique to describe how people lived there, Mithen perceptively injects colour into otherwise apparently inanimate and disconnected objects, while providing a feast of archaeological 'meat'.

Oppenheimer, S. *Out of Eden: The Peopling of the World* London: Constable, 2003.

A particularly valuable popular presentation of the revolution that has occurred in our thinking about the evolution of modern humans as a result of the advances in genetic mapping. This book is useful both because of being up to date and providing a particularly comprehensive presentation of the mapping results plus an exhaustive set of references. As such it is perhaps the best place to start to get to grips with the fascinating story of how humankind peopled the world.

Rudgley, R. *Lost Civilisations of the Stone Age* London: Century, 1998.

A thorough and intriguing exploration of the evidence for human intellectual accomplishments before the establishment of known civilisations. This book covers a range of important development from writing, art and science to mining, surgery and pyrotechnology.

Ryan, W. B. F. & Pitman, W. *Noah's Flood. The Scientific Discoveries About the Event that Changed History* New York: Simon and Schuster, 1998.

A stimulating book that provides a detailed introduction to the hypothesis that the biblical Flood was a folk memory of the catastrophic inundation of the shores of the Black Sea when the rising level of the Mediterranean Sea broke through the Bosphorus. Although this theory is now put in doubt by new geological evidence, this does not alter the fact that it is an useful introduction to many aspects of the 'Flood Myth'.

Sykes, B. *The Seven Daughters of Eve* London: Bantam Press, 2002.

An immensely informative description of the work behind the genetic mapping revolution of the 1990s. The description of both the science involved in genetic mapping and the thrill of discovery, although highly personalised, provides valuable insights into what we have found out about human evolution in recent years. The final chapters that seek to illustrate the personal lives of the seven daughters of Eve are, however, one-dimensional and a mite whimsical.

Wenke, R. J. *Patterns in Prehistory: Humankind's First Three Million Years*, 4th Edn. Oxford: Oxford University Press, 1999.

A highly readable and stimulating presentation that sets the scene for many of the social, political and economic issues addressed in this book. As such it is particularly valuable in assessing the potential impact of climate change on many aspects of human prehistory.

World Meteorological Organization (WMO). *Climate into the 21st Century*, Burroughs, W. J., ed. Cambridge: Cambridge University Press, 2003.

Drawing on an unrivalled selection of leading experts, this book provides a balanced and global coverage of climate issues. It identifies some of the most arresting examples of climate events; their impact on people's lives; and how we can face the future challenges of climate change.

Index